Chemical Biology

Chemical Biology

Applications and Techniques

Banafshé Larijani
Cancer Research UK

Colin. A. Rosser
Rye St Antony School, Oxford

Rudiger Woscholski
Faculty of Natural Sciences, Imperial College, London

John Wiley & Sons, Ltd

Copyright © 2006   John Wiley & Sons Ltd, The Atrium, Southern Gate, Chichester,
West Sussex PO19 8SQ, England

Telephone   (+44) 1243 779777

Email (for orders and customer service enquiries): cs-books@wiley.co.uk
Visit our Home Page on www.wileyeurope.com or www.wiley.com

Designations used by companies to distinguish their products are often claimed as trademarks. All brand
names and product names used in this book are trade names, service marks, trademarks or registered
trademarks of their respective owners. The Publisher is not associated with any product or vendor
mentioned in this book.

This publication is designed to provide accurate and authoritative information in regard to the subject
matter covered. It is sold on the understanding that the Publisher is not engaged in rendering
professional services. If professional advice or other expert assistance is required, the services of a
competent professional should be sought.

Other Wiley Editorial Offices

John Wiley & Sons Inc., 111 River Street, Hoboken, NJ 07030, USA

Jossey-Bass, 989 Market Street, San Francisco, CA 94103-1741, USA

Wiley-VCH Verlag GmbH, Boschstr. 12, D-69469 Weinheim, Germany

John Wiley & Sons Australia Ltd, 42 McDougall Street, Milton, Queensland 4064, Australia

John Wiley & Sons (Asia) Pte Ltd, 2 Clementi Loop #02-01, Jin Xing Distripark, Singapore 129809

John Wiley & Sons Canada Ltd, 6045 Freemont Blvd, Mississauga, Ontario, Canada L5R 4J3

Wiley also publishes its books in a variety of electronic formats. Some content that appears in print may not
be available in electronic books.

Library of Congress Cataloging-in-Publication Data

Chemical biology : applications and techniques / [edited by] Banafshé Larijani, Colin A. Rosser, Rudiger
Woscholski.
       p. ; cm.
    Includes bibliographical references and index.
    ISBN-13: 978-0-470-09064-0 (cloth : alk. paper)
    ISBN-10: 0-470-09064-2 (cloth : alk. paper)
    ISBN-13: 978-0-470-09065-7 (pbk. : alk. paper)
    ISBN-10: 0-470-09065-0 (pbk. : alk. paper)
   1. Biochemistry. 2. Biology. 3. Microscopy.   I. Larijani, Banafshé.   II. Rosser, Colin A.   III. Woscholski,
Rudiger.
   [DNLM: 1. Biochemistry–methods. 2. Lipids–chemistry. 3. Microscopy–methods. 4. Nuclear Magnetic
Resonance, Biomolecular–methods. QU 25 C5166 2006]
QP514.2.C4554 2006
612′.015–dc22   2006012747

British Library Cataloguing in Publication Data

A catalogue record for this book is available from the British Library

ISBN-13 978-0-470-09064-0 HB   ISBN-13 978-0-470-09065-7 PB
ISBN-10 0-470-09064-2 HB   ISBN-10 0-470-09065-0 PB

Typeset in 10.5/12.5pt Times by Thomson Digital
Printed and bound in Great Britain by Antony Rowe Ltd., Chippenham, Wiltshire
This book is printed on acid-free paper responsibly manufactured from sustainable forestry
in which at least two trees are planted for each one used for paper production.

Contents

5 Membrane potentials and membrane probes   67
Paul O'Shea

6 Identification and quantification of lipids using mass spectrometry   85
Trevor R. Pettitt and Michael J. O. Wakelam

7 Liquid-state NMR   95
Charlie Dickinson

8 Solid-state NMR in biomembranes   113
Erick J. Dufourc

9 Molecular dynamics

Michel Laguerre

10 Two-dimensional infrared studies of biomolecules

Xabier Coto, Ibón Iloro and José Luis R. Arrondo

11 Biological applications of single- and two-photon fluorescence

Banafshe Larijani and Angus Bain

Chapter 9 page: 133
Chapter 10 page: 151
Chapter 11 page: 163

12 Optical tweezers   199

Christopher Batters and Justin E. Molloy

13 PET imaging in chemical biology   217

Ramón Vilar

14 Chemical genetics   231

Piers Gaffney

Preface

"Chemical Biology" is a rapidly evolving interdisciplinary research area, where tools and techniques drawn from the physical and biomedical sciences are combined. Initially born out of a truly chemical synthesis approach to answer biological problems, other physical sciences disciplines such as physical and theoretical chemistry as well as physics, mathematics and engineering are becoming part of this new discipline. Thirteen different techniques and their corresponding biomedical applications drawn from the latter disciplines are portrayed in this modular textbook by leading physical investigators. Written for an audience with a basic understanding of the physical sciences this compendium is of particular value to biomedical researchers as well as advanced graduate students interested in venturing into Chemical Biology.

Given that Chemical Biology is in its infancy the editors are very conscious of the fact that no publication of this type can present an exhaustive documentation of currently available techniques. However it is hoped that the selected techniques and the corresponding examples for their application in biomedical research will serve to inform readers of their potential and inspire and encourage researchers to consider incorporating appropriately, physico-chemical methods in their research.

A website for this book can be found at www.wiley.com/go/larijanichemical

February 2006

Banafshé Larijani
Colin. A. Rosser
Rudiger Woscholski

List of Contributors

Stephanie Allen
School of Pharmacy
University Park
The University of Nottingham
NG7 2RD UK

Alicia Alonso
Unidad de Biofisca CSIC
Departamento de Bioquimica
Universidad del Pais Vasco
PO Box 644
Bilbao
Spain

José Luis R. Arrondo
Unidad de Biofisca, (Centro Mixto
   SCIC-UPV/EHU)
Departmento de Bioquimica
Universidad del Pais Vasco
PO Box 644 Bilbao
Spain

Angus Bain
Department of Physics & Astronomy
University College London
Gower Street London
WC1E 6BT UK

Christopher Batters
Division of Physical Biochemistry
MRC National Institute for Medical
   Research
The Ridgeway, Mill Hill
London NW7 1AA UK

Frank Booy
Department of Biological Sciences,
Faculty of Natural Sciences
Imperial College
Exhibition Road
London SW7 2AZ UK

Xabier Coto
Unidad de Biofisca, (Centro Mixto
   SCIC-UPV/EHU)
Departmento de Bioquimica
Universidad del Pais Vasco
PO Box 644 Bilbao
Spain

Charlie Dickinson
NMR Director
High Field Nuclear Magnetic Resonance
   Laboratory
Department of Polymer Science and
   Engineering
The University of Massachusetts-Amherst
Amherst MA 01003
USA

Erick J. Dufourc
Diracteur de Recherche CNRS
UMR-CNRs 5144 MOBIOS
Institute Europeen de Chimie et Biologie
2 rue Robert Escarpit
33607 Pessac
France

Piers Gaffney
Room 306, RSC1
Department of Chemistry
Imperial College of Science and
   Technology and Medicine
Exhibition Road
London SW7 2AZ UK

Félix Goñi
Unidad de Biofisca CSIC
Departamento de Bioquimica
Universidad del Pais Vasco
PO Box 644
Bilbao
Spain

Ibón Iloro
Unidad de Biofisca, (Centro Mixto
   SCIC-UPV/EHU)
Departmento de Bioquimica
Universidad del Pais Vasco
PO Box 644 Bilbao
Spain

Michel Laguerre
Chef de Project
UMR-CNRs 5144 MOBIOS
Institute Europeen de Chimie
   et Biologie
3 rue Robert Escarpit
33608 Pessac
France

Banafshé Larijani
Head of Laboratory
Cancer Research UK
Cell Biophysics London Research
   Institute
44 Lincoln's Inn Fields
London NW7 1AA UK

Justin E. Molloy
Division of Physical Biochemistry
MRC National Institute for Medical
   Research
The Ridgeway, Mill Hill
London NW7 1AA UK

Paul O'Shea
Cell Biophysics Group
The School of Biomedical Sciences
The University of Nottingham
NG7 2UH UK

Trevor R Pettitt
Cancer Research UK
CR-UK Institute for Cancer Studies
Birmingham University
Birmingham, B15 2TT UK

Colin. A. Rosser
Director of Studies
Rye St Antony School
Oxford UK

Ramón Vilar
Instittucio Catalana de la Reserca I Estudis
   Avancats (ICREA)
43007 Tarragona
Spain

Michael J.O. Wakelam
Cancer Research UK
CR-UK Institute for Cancer Studies
Birmingham University
Birmingham, B15 2TT UK

Rudiger Woscholski
Division of Cell and Molecular Biology
Faculty of Natural Sciences and Chemical
   Biology Centre (CBC)
Imperial College
Exhibition Road
London SW7 2A2 UK

1

Introduction

Banafshé Larijani,[1] **Colin. A. Rosser**[2] and **Rudiger Woscholski**[3]

[1]*Cancer Research UK, London Research Institute, 44 Lincoln's Inn Fields, London, UK*
[2]*Director of Studies, Rye St Antony School, Oxford*
[3]*Division of Cell and Molecular Biology, Faculty of Natural Sciences and Chemical, Imperial College, London*

1.1   Chemical biology – the present

The phrase 'chemical biology' is gaining momentum in the literature, which is documented by the increasing number of journals dedicated to this new discipline. The fact that there is now a journal *Nature Chemical Biology* speaks for itself. However, like many new developments in science chemical biology is adjusting itself and while doing so is opening up to other disciplines and subdisciplines. Born out of a truly (and for that reason synthetic) chemical approach towards unravelling biological problems and systems, chemical biology was initially employed for high-throughput screening of small molecules using biological (cellular) assays. Now, in the post-genomic era, the manipulations of proteins and DNA by small molecules have been added to the arsenal of chemical biology techniques. This compromises the generation of chemical receptors or catalytic antibodies, the expansion of the genetic code with unnatural amino acids and the creation of the new field of chemical genetics, to name a few. While most of these techniques are still very much relying on the synthetic organic chemistry that initiated chemical biology, there is also a trend to widen the applications and in doing so to incorporate other disciplines. Structural biology, a discipline that is very much influenced by physical scientists, is for many of these applications an important provider of the necessary data. Furthermore, system-wide approaches that rely on the processing of vast amounts of data will need mathematical skills. This opening up to other disciplines such as physicists and mathematicians is adding a 'quantitative' dimension towards the originally qualitative chemical biology.

Chemical Biology   Edited by Banafshé Larijani, Colin. A. Rosser and Rudiger Woscholski
© 2006 John Wiley & Sons, Ltd

Molecular and cellular biology research of the past 20 years was in principle more qualitative than quantitative. For example, the majority of cellular signalling and membrane traffic investigations were aimed at describing/mapping the flow of information (cellular signalling) or cargo (membrane traffic) in the cellular context. This research, together with other disciplines such as developmental research, was very successful in monitoring events in the living cell, thus fostering a demand for new technologies with *in vivo* capabilities. Imaging technologies such as FRET/FLIM, which now combine suitable spatial resolution with the ability to collect data in a quantified fashion, open up the cellular space for the determination of physico-chemical parameters such as binding affinities between two distinct proteins. The aim for quantification is a consequence of the successful (and thus sufficiently deep) mapping of the qualitative network of chemical reactions (e.g. enzymes) and interactions (e.g. receptors), which have accumulated enough data to increase the desire to understand the underlining principles governing the dynamic behaviour of these networks (e.g. systems biology). It seems that biological research is going through a cycle were a wealth of qualitative data will create a demand for a quantitative assessment, which in turn will foster the development of new techniques, tools and methods to determine the physico-chemical properties and dynamics of metabolites and proteins in the living cell. The focus on the 'living' cell as the ultimate aim in biological research is probably were the metaphor of a cyclic history is most distant. While the 'founding' era in biochemistry was interdisciplinary and quantitative, it lacked most techniques and tools to investigate the living cell. This 'live' dimension distinguishes chemical biology from 'biochemistry/biological chemistry', the latter being otherwise as interdisciplinary and quantitative. Therefore, chemical biology is not a simple re-branding of 'biological chemistry/biochemistry', but could probably be seen as a delayed/paused development of the latter, an offspring of biochemistry.

1.2 Chemical biology – the past

The physical sciences, in particular physics and chemistry, were of considerable importance for the development of the then new discipline 'biochemistry'. Not surprisingly, physicists or chemists working with biological materials and systems were responsible for many important discoveries in the field of biochemistry in the past. At that time the emerging biochemistry discipline was truly benefiting from the cross-fertilization of the contributing disciplines biology, chemistry and physics. The early 'biochemists' were, and actually had to be, genuinely interdisciplinary in their approach, controlling the biological materials and systems with physico-chemical techniques. A good example of an early biochemistry development might be the assessment of the enzymological properties, like K_M (Michaelis–Menten constant) and V_{max} (maximum reaction rate). Enzymes have been around for a long time and their 'fermentation' properties were of considerable commercial interest (for breweries, etc.), which probably aided research into their properties and mechanisms. Since an enzyme is a catalyst for a given chemical reaction that uses some starting material (the substrates) to create the corresponding products, the parallels with chemical synthesis and analysis are quite obvious. Monitoring an enzyme reaction accurately in a crude extract was an essential prerequisite for its purification and subsequent molecular characterization. The quantification of the disappearance of substrates and/or appearance of products in the shortest possible time that can be measured was a vital ingredient in the early enzymological research. Achieving this goal was only possible by employing physical methods (i.e. stopped flow technique) as well as chemical analytical techniques (i.e. colorimetric measurements of products or substrates). In more complex cases this involved the separation/purification of the substrates and/or products, usually by chromatography, or the connection to a well-characterized enzyme, which was used to transform a given product towards one that can be monitored. Employing these assays led to a wealth of information about the enzymological properties, which was complemented by the theoretical work of many investigators, such as Michaelis and Menten, providing mathematical frameworks for the observed turnover by the biological catalysts, the enzymes. Understanding the mechanism of enzymes was another area that profited from cross-fertilization of the biological and physical sciences. In the absence of any molecular biology to obtain bulk amounts of proteins, most advances at the time were limited to very abundant enzymes, which possessed an inherent robustness and stability, allowing their corresponding purification (a process that could take up months or years) and subsequent analysis with respect to their molecular and physico-chemical properties. Inhibitors, activators, pseudo-substrates and substrate-mimetic compounds were employed together with pulse-chase labelling techniques to unravel enzymological mechanisms. Again, synthetic and analytical chemistry were needed as much as physical techniques to gain the knowledge that is now confined to textbooks for both schools and universities. Mechanisms were then re-enacted using synthetic organic chemistry to provide the final proof for the deducted mechanisms, an approach that was most prominent in the field of 'bioorganic chemistry'. Thus, it seems that enzymology is a truly interdisciplinary and more importantly a truly quantitative area within the biochemistry discipline, which can easily be recognized as fitting into current concepts of 'chemical biology'. However, the current enzymological technologies, which were all developed in the past, are still rarely employable in the living cell. As pointed out above, chemical biology is aiming to provide tools and techniques that will be ultimately applicable to living cells; it is this goal that drives 'chemical biology' towards an area that 'biological chemistry/biochemistry' could not cater for. Thus, chemical biology could be the natural successor to biological chemistry/biochemistry, taking on their heritage and adding a new cellular dimension.

1.3 Chemical biology – the future

While biologists participated strongly in the formation of this new discipline, their most influential time came with the ascent of molecular biology. Molecular biology was instrumental in developing the sequencing techniques that were ultimately responsible for the mapping of the genomes of whole organisms. This led to a wealth of data and to a new era aptly named 'post-genomic'. This new area is characterized on one side by the desire to obtain a comparable wealth of data for the other components in a living cell, the proteins and metabolites (proteomics and metabolomics). On the other hand there is an increasing desire to get a 'quantitative' grip on the 'omics' data as well as any other biological problem that has been so far investigated in a qualitative (and thus descriptive) fashion. Out of this desire at least one new profession has been created, bioinformatics. Employing mathematical skills and techniques to organize, analyse and ultimately predict biological data is truly needed. However, bioinformatics and other theoretical approaches are at present not yet in a position to provide the necessary depth of simulation and prediction due to a limitation in processing power and even more importantly the lack of genuine 'quantitative' data. The latter has to come from an interdisciplinary approach applying the methods and techniques of the physical sciences to biological problems and systems. Systems biology, another recent development, could be easily classified as 'chemical biology'. However, if a more catholic definition is applied to chemical biology (pure synthetic chemistry applied to biological problem-solving), then systems biology could be seen as a natural sister discipline that aims towards quantification of biological networks using mathematical, physical and chemical methods. One could easily see that both systems biology and chemical biology will be competing for dominance in years to come. Both could claim to make use of the other, consigning it as an aiding discipline. It remains to be seen whether chemical biology will absorb the other offshoots and become, like biochemistry, a genuine independent discipline. Early indications support this notion, since many departments already carry the name 'chemical biology', giving chemical biology a head start over others, such as systems biology.

1.4 Chemical biology – mind the interdisciplinary gap

The interdisciplinary character of chemical biology is a blessing at the best of times, but could be a curse in the worst-case scenario. While the advantages of any multidisciplinary setting are obvious, implementing and completing such an undertaking could be quite laborious and exhausting to all participating partners. There is a 'cultural' divide that is made most obvious by different meanings of common words and phrases. While most biologists would anticipate that the words 'system' or 'energy' have quite distinct meanings from those used in the physical sciences, they are less likely to expect similar differences in meaning from words such as 'model' or 'molecular'. While these linguistic differences will be overcome quickly once the talking starts, they are still indicators of a different culture in approaching research, for which there should be an increased awareness.

Interdisciplinary research has barriers that need to be crossed. Biological understanding by physical scientist and, vice versa, physical understanding by biological scientists will not be sufficiently present to enable a flying start. This knowledge barrier is not insurmountable but is nevertheless hindering or delaying progress. It is thus essential to provide potential interdisciplinary users with some guidance about the potential techniques the physical sciences can offer. Given that there is much already in the literature concerning the 'traditional' chemical synthesis-involving approach, we focus here on the more physical dimension that has recently emerged and try to showcase a selection of interesting techniques, presented by physical scientists to entice biologists.

1.5 An introduction to the following chapters

The following chapters have been contributed by a set of international scientists actively developing a range of potential techniques that the physical sciences have to offer. In the subsections below is a brief summary of each chapter's content.

1.5.1 Cryo-electron microscopy

Cryo-electron microscopy is a leading method of structural investigation in biology, providing a direct visualization of the specimen of interest in a near-native state with structural preservation to atomic resolution. A thin film of the native sample is quench-frozen in liquid ethane so that it is transformed into an amorphous 'glassy' state prior to transfer to and imaging in the electron microscope, still frozen, under conditions of low electron exposure. The low contrast images that result can be enhanced significantly by image processing. Chapter 2 reviews the reasons for needing such electron microscopy and shows some of the highlights achieved to date. With software developed during the last decade the technique can be used to obtain structures of molecules in different conformational states with a resolution of ~ 10Å. The ability to examine whole cells that is emerging with tomographic analysis is bringing the dream of understanding the dynamics of cells in terms of the function and localization of proteins, their complexes and their relationship to organelles a step closer. Proteomics and systems biology are here with electron microscopy.

1.5.2 Atomic force microscopy

Chapter 3 starts with a brief history of microscopy complementing Chapter 2 and then introduces atomic force microscopy (AFM), which belongs to a class of techniques known as scanning probe microscopy (SPM). Instead of using light, it uses a sharp probe that touches and scans a sample's surface to gather topographic image information. During imaging, a piezoelectric scanner accurately controls the position of a cantilever in three dimensions and an optical detection system is used to amplify the tiny deflections of the cantilever. These signals are fed back to a computer to generate a three-dimensional image of the sample's surface topography. AFM is especially useful for observing biological specimens as it allows the determination of surface properties in a variety of media including air and liquid. Hence AFM can provide information on biomolecular structure and biological processes and in recent years AFM has been developed to visualize dynamic biological processes with video-rate imaging. In addition to imaging, AFM is able to record forces acting between the probe and the surface. This can be exploited to explore, to the level of a single molecule, the dynamic strengths of biological receptor–ligand bonds and the mechanical properties of biopolymers, including proteins, polysaccharides and nucleic acids. The use of AFM in measuring membrane surface potentials is discussed in Chapter 5.

1.5.3 Differential scanning calorimetry in the study of lipid structures

In this technique two containers, one with a sample and the other used as a control, are identically heated electrically. If a phase change occurs in the sample, when for example it reaches its melting point, then the energy supplied is used to melt the sample. During the melting process the sample container remains at the same temperature, whilst the control container continues to heat up. Thus the technique can measure the energy difference, i.e. the energy required for the phase change, and record the temperature at which it occurs. When used with bio-molecules, differential scanning calorimetry (DSC) allows the measurement of the minute exchanges of heat (in the microcalory range) that accompany certain physical processes, such as aggregation/dissociation, conformational changes or ligand binding. DSC does not perturb the system with chemical probes, is often non-destructive, and can nowadays be performed with very small amounts of material, i.e. tens of micrograms. Typical applications of DSC include protein studies such as thermal stability and thermal denaturation. Chapter 4 considers the use of DSC in the field of lipids.

1.5.4 Membrane potentials and membrane probes

The fluid-mosaic model of cellular membranes has helped enormously in devising further experimental investigations of the properties of membranes. Much basic knowledge was accumulated throughout the 1970s and 1980s as well as an enormous expansion of information concerning how signalling reactions take place at and within membranes. A commonly held view of (perhaps) the sole function of membranes, however, is that they act to compartmentalize soluble components (from ions to enzymes) of cells by acting simply as a selective permeability barrier. However, towards the end of the 1990s it was appreciated that molecular (and particularly macro-molecular) diffusion within membranes was not as clear-cut as first envisaged, and this pointed to a more sophisticated membrane organization. Membranes possess properties that underlie subtle and highly sophisticated additional modes of behaviour. These include a repertoire of different electrical potentials and indicate that in many ways the membrane represents a unique molecular environment in its own right. In Chapter 5 these electrical properties are outlined, their key roles are suggested and methods of measurement are discussed.

1.5.5 Identification and quantification of lipids using mass spectroscopy

Mass spectroscopy (MS) is useful for determining exact structures based on characteristic fragmentation patterns. It usually requires an essentially pure compound, otherwise, as in unresolved lipid samples composed of many closely related structures, the complex fragmentation patterns become largely uninterpretable. Chapter 6 describes the technological advances of both the separation techniques and the mass spectrometers as well as the way in which these have been combined. High-pressure liquid chromatography (HPLC) has increased the sensitivity of separation, and apparatus for this technique can be found coupled to a mass spectrometer. Advances in MS technology have produced fragmentation techniques that are less harsh and electrospray ionization mass spectrometry (ESI-MS), together with tandem mass spectrometry (MS/MS), has made the precise identification and quantitation of global lipid species a feasible aim.

1.5.6 Liquid-state NMR

Nuclear magnetic resonance (NMR) is a pre-eminent technique for molecular-level understanding because it exquisitely displays differences in and connections between chemical environments of atoms in a molecule. Many biologically relevant nuclei – ^{1}H, ^{13}C, ^{31}P, ^{15}N, ^{14}N, ^{2}H, ^{23}Na, ^{17}O ^{19}F and more – can be usefully observed by NMR and only a few micrograms are sufficient for proton NMR work. In the past 25 years with vastly improved sensitivity and sophistication of imaging capability, numerous areas of the biological sciences have taken advantage of the power of NMR to observe selected components of organisms, both *in vivo* and *in vitro*. There are myriad biological applications of NMR to almost every tissue and fluid of animals and plants already in the literature. In Chapter 7 a few examples have been given in order to illustrate the potential of NMR to supply salient details to almost any area of biological research. The use of NMR to monitor certain processes *in vivo* and identify components of complex mixtures is exemplified. Also, as NMR has frequently been used as a structural tool, and indeed complete three-dimensional structures of proteins up to and over 20 kD a molecular weight have been determined in solution, a representative structural study of a peptide is presented.

1.5.7 Solid-state NMR in biomembranes

Chapter 8 complements the previous chapter and develops the basic theory needed to understand how solid-state NMR can be utilized in membrane systems. In particular it shows that wide line NMR can be sensitive to molecular motions and that it is an appropriate tool to track membrane dynamics (membrane fluidity, fusion) and probe average orientations of molecules embedded in membranes (membrane topology). The three-dimensional structure of molecules in membranes can also be obtained by making use of magic angle sample spinning, a technique that leads to pseudo 'high-resolution' spectra as in the liquid state, the sample being still in the membrane 'liquid crystalline' state. Another facet of NMR is also presented, i.e. relaxation studies, that allow measurements of the speed of molecular motion, activation energies and the picturing of membrane dynamics from the atomic level where intra-molecular motions dominate to the cell level where membrane hydrodynamic modes of motion play an important role. Selected examples are discussed: lipid phases, effect of amphipathic peptides, bilayer internal structure and dynamics, effect of sterols on membrane fluidity and thickness, orientation of sterols and peptides in membranes, three-dimensional structure of molecules in membranes or colloidal aggregates and topology of peptides in membranes.

1.5.8   Molecular dynamics

In this chapter the theory of molecular mechanics and molecular dynamics is presented to provide the background for computer modelling of biological systems. The basis of a computer simulation is explained using examples such as the building up of a lipid bilayer and embedding of a protein or peptide within it, and results of such simulations are shown. The technique gives a valuable method for exploring the conformational space of a system and how molecular fragments might behave dynamically at an atomic level. One of the major advantages of this technique is that the predictions produced can be used to inform further experimental investigations.

1.5.9   Two-dimensional infrared studies of biomolecules

This chapter describes a new and developing form of infrared spectroscopy that uses two ultrafast IR laser pulses of different frequencies to examine biomolecules using experimental methodology similar to that developed for NMR. It is proving to be a very powerful technique to obtain information about the structural features of complex molecular systems in solution as a function of time. The technique can provide structural and dynamic information about proteins and peptides that cannot be obtained from linear spectra. It allows the measurement of coupling and angular relations between pairs of modes, for example the N–H and C=O modes of peptides. In addition the effect of a perturbation to the system such as a change in pH or temperature can be investigated with the resulting spectra, providing detailed information about the time-dependant behaviour and interaction of these molecules from changes in the bond stretching frequencies. Order/disorder transitions and hydration states can be identified using this technique.

1.5.10   Biological applications of single and two-photon fluorescence

Chapter 9 describes how single photon and two-photon fluorescence may be exploited in biology and more specifically in molecular signalling. Fluorescence has been exploited as a tool for investigating structure and the dynamics of molecules in their microenvironment because fluorescent molecules are highly sensitive to their surroundings. The principles of resonance energy transfer (RET) by single photon excitation are explained and the fluorescence imaging methods (FLIM) used to quantify and localize protein–protein and protein–lipid associations in cells are described. Two-photon fluorescence, which reduces photo damage, allowing longer periods of data acquisition, is then described and exemplified as both a contributor to fluorescence imaging but also a less invasive method of scanning confocal microscopy.

1.5.11 Optical tweezers

The purpose of this chapter is to give the reader a practical introduction to optical tweezers; how they work, how they are built, how they can be used to make calibrated measurements on single molecules and how a paradigm system like actin and myosin can be studied. A future challenge is to combine the use of other single molecule techniques with optical tweezers so that parallel measurements can be made on the same single molecule. Such studies will give detailed insights into enzyme mechanism. One would also like to make many simultaneous measurements from multiple, individual molecules within a living cell so that we can build up a picture of how proteins, DNA and ligands interact in intact systems.

1.5.12 PET imaging in chemical biology

Monitoring molecular events *in vivo* and in real time is essential to gain a better fundamental understanding of biochemical processes, to develop novel means of early detection of disease and to design new drugs. Consequently, there has been great interest in finding suitable techniques with the appropriate sensitivity, selectivity and spatial resolution to be able to image biological processes at a molecular level.

Positron emission tomography (PET) and single-photon emission computed tomography (SPECT) have the ability to monitor metabolic processes *in vivo*. These techniques make use of highly sensitive imaging agents or reporter probes that can be designed to selectively interact with specific tissues or biomolecules, providing a much more detailed picture of the targeted structure or biological processes at a molecular level. Chapter 11 introduces positron emission tomography (PET) as a molecular imaging modality and provides specific examples where PET has been successfully used to get a better understanding of specific biological processes, in the detection of disease and also as a tool for drug discovery and development.

1.5.13 Chemical genetics

In this chapter, chemical genetics, defined as the search for chemical agents that act as conditional switches for gene products, either inducing or suppressing a state or phenotype of interest, is explored. Here it is suggested that chemical genetics has the potential to offer insights into biological systems as powerful as, and somewhat complementary to, classical genetics but warns that to realize its potential this will require fully integrated interdisciplinary scientific endeavour: sophisticated organic synthetic skills, a broad appreciation of the biological sciences, and investment in technologies such as robotics and image analysis. The need for molecular libraries is identified as an essential requirement for future development in this area.

2
Cryomicroscopy

Frank Booy and **Elena Orlova**

Imperial College London, London, UK

2.1 The need for (electron) microscopy

A light microscope extends the range of the eye, enabling visualization of objects otherwise too small to be seen. Ultimately its resolving power is limited to ~1 μm by the wavelength of the light used. This means that we can see a bacterium but not a virus particle. To extend the eye's range further we must use radiation of not only a shorter wavelength but also a type for which we can make a focussing device (lens) to form an image. X-rays have a shorter wavelength (e.g. Cu $K_\alpha \approx$ 1.5 Å) but no simple lens is available. Electrons are charged particles (waves); at 100 kV their wavelength is 0.037 Å and in a magnetic or electric field they experience a force and can thus be focussed. Therefore an electron microscope (EM) is simply a microscope that uses electrons instead of light. The first such microscope was developed by Ruska in the 1930s (Nobel Laureate 1986), but its widespread application and development was delayed by the second World War. In fact further technological development was required before it was to become useful for examining frozen, hydrated specimens.

2.2 Development of cryomicroscopy

Cryo-electron microscopy did not come of age until the 1980s, after the development of differential pumping permitted film to be separated from the frozen sample by a factor of 100 or so in vacuum in the microscope, improved anti-contamination systems prevented the sample itself acting as a cold trap and stable specimen cooling

---

Chemical Biology Edited by Banafshé Larijani, Colin. A. Rosser and Rudiger Woscholski
© 2006 John Wiley & Sons, Ltd

holders became available. Gatan developed the most widely used routine side-entry cooling holder. Siemens developed the ultimate holder with stability achieved by cooling the complete specimen and objective lens assembly to 4 K using liquid helium. As a consequence the lens became superconducting. Unfortunately Siemens closed their electron optical division and it was left to the Berlin group (under Zeitler) with Philips (now FEI) to develop this into a fully functional cryo-EM (due largely to the dedication of Zeitler and Zemlin). This functioned beautifully until retirement closed the group in 1997. The microscope has now been moved to Goettingen, Germany. Fujiyoshi and collaborators in Japan developed a very stable, top-entry (and thus fixed-tilt) cryo-EM in the 1990s running at liquid helium temperature (4 K) (now marketed by Jeol), and FEI developed a eucentric, helium-cooled cryo-EM. These are the more common top-end cryo-EMs of today, capable of atomic resolution augmented by their use of 200–300 kV and field emission guns.

A modern electron microscope is capable of ~1 Å resolution and is shown schematically in Figure 2.1. It consists of a source of electrons, a number of lenses,

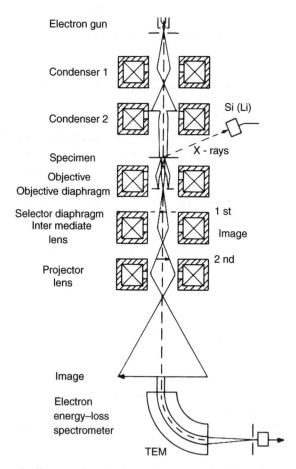

Figure 2.1 Schematic diagram of an electron microscope with X-ray detector and post column energy filter. [Reproduced from Hawkes (2004)]

a viewing screen and a data collection system. The simplest electron source is a heated tungsten filament. When electrons have energy greater than the work function of tungsten, they escape and form a pool of free electrons that may be accelerated in an electric field. If a crystal, commonly LaB_6, is attached to the filament, electrons may be induced to emit from a single crystal face and emerge with greater coherence and current density from a smaller, effective source. A further improvement occurs when the electrons are extracted by field emission from a tungsten crystal, and this is the source of choice for cryo-EM. The tip may either be cold or partially heated (Schottky emission). The electrons are then accelerated through voltages typically of 100–300 kV when they are travelling at relativistic speeds ($\sim c/2$ at 100 kV). The convergence and energy density of the electron beam are controlled by two condensor lenses prior to interaction with the specimen. The objective lens is the contrast-forming lens with an aperture in its back focal plane and provides a basic increase in magnification ($\times 20$–50). The image formed by this lens is subsequently magnified by a further series of intermediate and projector lenses before the electrons impinge on a fluorescent viewing screen, CCD (charge-coupled device) camera detector or sheet of film. It is also possible to image the back focal plane of the objective lens, thus forming an electron diffraction pattern.

2.3 Sample–electron interaction

Electrons have a large scattering cross section compared with, say, X-rays and interact very strongly with matter. For this reason an electron microscope has to be maintained under vacuum and only dry samples can be investigated. This has serious consequences for biological materials, which comprise typically some 85 per cent water. Thus tremendous effort has gone into developing methods for supporting samples as they are dehydrated prior to observation. Methods involving chemical cross-linking and staining ultimately distort the native structure and thus limit the details visualized and the resolution obtained. Sucrose, glucose, trehalose or tannic acid have been used to stabilize/replace bound water both at room temperature and in the frozen, hydrated state. However, water is unique and nothing can replace it without introducing potentially significant, structural changes. An alternative approach to dehydration was suggested by Fernandes-Moran, namely to look at samples maintained in a frozen state inside the microscope. At a low enough temperature, the vapour pressure of water is sufficiently small for the sample to retain its water. The viability of cryo-EM was demonstrated by Chanzy in Europe and Glaesar and Taylor in the USA in the 1970s, where electron diffraction experiments showed that freezing could preserve high-resolution structure in hydrated polysaccharide and protein crystals. Also in the 1970s, Parsons developed an environmental cell in which the sample was maintained at room temperature in a fully hydrated state and he was able to demonstrate high-resolution structural preservation by electron diffraction. However, a consequence of the strong interaction of electrons is that the thickness of the sample that may be investigated is limited and thus the wet cell, with its associated water vapour, was impractical for imaging.

Electrons may be scattered elastically (with no loss of energy) or inelastically (with loss of energy). Inelastically scattered electrons are not brought into focus by the objective lens in the same plane as elastically scattered electrons. They can, however, provide useful information about the atomic make-up of the sample in energy loss spectroscopy. With 100 kV electrons the mean free path in ice (the average distance the electron travels before interaction) is $\sim$1000 Å. For an interpretable image this is an indication of appropriate maximum specimen thickness. This increases for more energetic electrons at higher accelerating voltages. Typically, as the thickness of the sample increases, the resolution that can be achieved decreases due to plural and inelastic scattering. For these reasons high-resolution studies by cryo-electron microscopy have been limited to thinner samples. The use of an energy filter to remove inelastically scattered electrons from the image does enable somewhat thicker samples to be investigated and is currently important in tomographic studies. However, thicker samples, such as whole cells, must be first sectioned or fractured before microscopy, and their treatment is beyond the scope of this chapter.

2.4   Contrast in negatively stained and cryo preparations

Prior to the advent of cryo-EM, the standard method for examining smaller specimens was negative staining, first introduced by Ruska in 1937. Indeed, staining remains the simplest method for specimen observation to $\sim$20 Å. In this technique the sample is dried down in a pool of a heavy metal salt (typically uranyl acetate) that supports the structure and maps out its surface features. Strong image contrast results from the scattering difference between the heavy metal surrounding the structure and the protein, which is seen as a light object in a darker background.

The basic idea of cryo EM is very simple and is illustrated in Figure 2.2. A drop of the specimen of interest is applied to a continuous or holey or lacey carbon film on an EM grid, reduced in thickness by blotting with filter paper and then plunged into liquid ethane maintained close to its freezing point by liquid ethane. Supercooling takes place at some 10^4–10^6 °C/s resulting in vitrification of a thin specimen in the µs to ms time range. Excess ethane is then blotted off and the grid is transferred into a specimen cooling holder for examination and photography close to liquid nitrogen temperature. Refinements to freezing equipment now permit the control of parameters such as humidity and temperature prior to freezing, automated, reproducible blotting and the release of caged molecules or spraying of molecules to induce defined changes just prior to freezing.

Because the sample is chemically untreated it is acutely sensitive and susceptible to damage by the beam of ionizing radiation (much as we would be!) and must be imaged under the lowest dose conditions that are compatible with an acceptable signal-to-noise ratio. To a first approximation, the scattering of electrons is proportional to the density of the material. Thus a protein, with a density of $\sim$1.3, is seen as a darker object in the less dense aqueous background (density $\sim$0.95 for ice), which means that the contrast has been reversed as compared with visualization in negative stain. Figure 2.3 shows T4 phage as an example. The contrast obtained is low and normally enhanced by defocusing the objective lens.

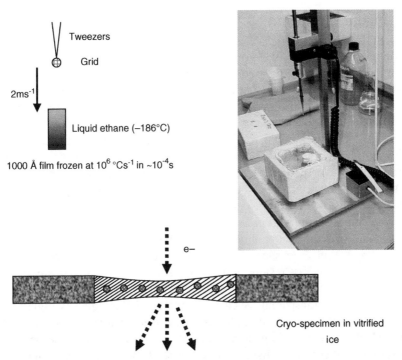

Figure 2.2   Left: schematic diagram to show that plunging a grid at 2 m/s into liquid ethane cooled by liquid nitrogen achieves a vitreous state for a 1000 Å layer in the ms to μs time frame with surface cooling rates $\sim 10^6\,°Cs^{-1}$. Right: traditional freezing equipment. Below: schematic diagram of vitreous sample spanning a hole in a grid in the EM

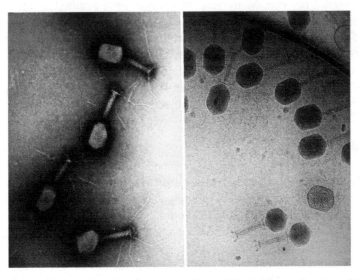

Figure 2.3   The appearance of T4 phage after negative staining with uranyl acetate (a) and in the frozen, hydrated state (b). Note the reversal of contrast between (a) and (b) [Micrograph (a) is courtesy of Dr Naiqian Cheng]

2.5 Image formation

Electron microscopes are able to provide images at very high magnification. There is, however, a trade-off between high and low magnification. A microscope is designed to give its optimum electron-optical performance at high magnification but the electron dose (for the same optical density on the film) depends on the magnification squared. It is thus of acute interest to know at how low a magnification a particular EM will still provide atomic resolution. Often the limitation is a stray magnetic field. A further complication is that the images obtained do not provide a direct representation of the specimen spatial density distribution. The contrast results from the scattering differences between the protein and water modified by the properties of the objective lens. Image contrast arises from two components: amplitude and phase. Amplitude contrast results from absorption or strong scattering of the incident beam by the sample with the scattered electrons removed from the image by the objective aperture. For a thin cryo specimen the amplitude contrast has been estimated at 7 per cent. Phase contrast results from interference of elastically scattered electrons with the unscattered beam; the amplitude of the scattered wave is unaffected. It is as if the object were transparent. Such thin, unstained, biological samples that only scatter electrons through small angles are often described as 'weak phase objects'. Additional phase contrast is generally introduced by defocusing the objective lens when spherical aberration enhances the contrast of images further. A more detailed description of image formation can be found in Reimer (1997). The visualization of weak phase objects by defocusing results in considerable complexity in interpreting the image as a simple projection of the specimen. This arises because image details in a particular size-range may be present in the image with different contrast, even complete contrast reversal. Many years of analysis have resulted in the development of specialized software to restore the original density distribution of images. This is referred to as correction of the contrast transfer function of the EM (or simply CTF correction). After this correction, the image of the weak phase object can be considered as a true two-dimensional projection of the three-dimensional object. This permits the use of special mathematical approaches to obtain a complete three-dimensional density distribution for the object.

2.6 Image analysis

The development of specialist software for the analysis of electron micrographs has equipped researchers with a variety of computational tools to analyse different types of sample. These methods are all based on the premise that a micrograph is a simple projection of the object and therefore have much in common. The main steps include (i) pre-processing of images, (ii) restoration of images, (iii) enhancement of images, (iv) determination of orientations and (v) reconstruction of the three-dimensional distribution of density. The result obtained must then be validated and interpreted.

Pre-processing is a set of operations designed to transform the data into the format required for the software, to determine the defocus at which the micrograph was

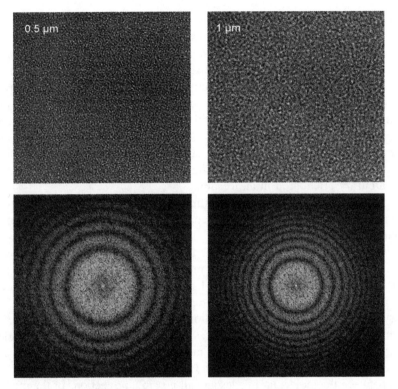

Figure 2.4 Images of a carbon film (taken at 200 kV on an FEI T-20 FEG-EM) at the defocus values indicated and their corresponding Fourier transforms. The light and dark bands are Thon rings which result from the contrast transfer function imposed by the objective lens of the microscope

taken and to normalize the data. Band pass filtering is usually applied to remove any uneven background and unimportant high-frequency detail.

The *restoration* step aims to restore the frequency spectrum of the image. Before the more common availability of electron microscopes equipped with a field emission gun, few image details were present in micrographs beyond the first Thon ring and this step was less important. Today, a typical high-quality, cryo-micrograph might have a first zero of the CTF at ~18 Å and information present to 6–10 Å in subsequent Thon rings (see Figure 2.4). To analyse such data beyond the first Thon ring requires careful correction of the CTF. In the most basic approach, this is achieved by simply reversing the phase of each odd Thon ring; a more complete approach requires additional correction of the amplitudes. This modifies the contrast of image details so that the image then reliably represents a true projection of the three-dimensional object and may be used to generate a more faithful reconstruction.

Enhancement is the procedure used to increase the signal-to-noise (S/N) ratio of images by averaging. For all types of specimen, there will be local variations in the thickness of the ice film, in concentration of the buffer salts and other contaminants and impurities such as denatured proteins. Random noise variation also arises from the support film and the effects of radiation damage on the ensemble. If micrographs are recorded on film, the photographic emulsion will add shot noise; a CCD camera/photo multiplier also produces some additional noise. Although crystals and nonperiodic specimens are treated differently, the essence is simply averaging of similar particle images. For two-dimensional crystals averaging can be done using the Fourier spectrum of the crystal by extracting the periodic part of the spectrum and suppressing the background. The reverse Fourier transform then provides an average image in which the unit cell of the crystal is seen with an enhanced S/N ratio. For single particles, averaging is performed in real space, where the images are first aligned and then classified into different groups according to their various features. The assumption is made that the sample comprises identical subunits, each embedded with an arbitrary orientation in an ice film.

Determination of the orientations of different projections is an essential step prior to calculating a full three-dimensional map. In some cases, some information can be obtained during data collection, e.g. the tilt of the goniometer at which the images were recorded can be used initially in the tomographic analysis of two-dimensional crystals. For single particles the task is much more difficult. Today, there are two principal approaches, common lines and projection matching. The common lines approach is based on the fact that any two, 2-dimensional projections of a structure have at least one, 1-dimensional projection in common. Knowledge of the angles between these lines for at least three, 2-dimensional projections provides the necessary orientational parameters. The method was first used in reciprocal space by Crowther and later in real space by van Heel. In projection matching a comparison is made between the molecular images and all possible projections of the model. The orientation of the projection that has the closest similarity to the molecular image is assigned to the image. This approach is widely used in real space for particles with low symmetry and in Fourier space for particles with icosahedral symmetry.

In *reconstruction* of the three-dimensional distribution of molecular densities, there are two main approaches. The first is a real space approach and the second a Fourier space approach that is analogous to the crystallographic method. In real space, the 'back projection' technique is used to reverse the operation of obtaining a projection. A projection simply represents the total sum of all densities of the three-dimensional object in a single plane (somewhat like a medical X-ray). To restore the densities of the three-dimensional object the densities of the projections must be extended in the reverse of the projecting direction. There are several algorithms that perform this procedure.

The Fourier method is based on the central section theorem, which states that the Fourier transform of a projection is a central section in Fourier space. This means that projections at different angles then provide sections of Fourier space at these angles and thus the space can be filled up. We can thus obtain the complete three-dimensional Fourier transform of the object. The reverse Fourier transformation of such a volume will generate the three-dimensional density distribution of the object in real space. For particles with icosahedral or helical symmetry, a Fourier–Bessel transformation is widely used since the use of a cylindrical coordinate system may avoid some interpolation errors.

Validation and interpretation of the results obtained are necessary so that the quality of the reconstruction that has been obtained can be assessed. This can be analysed computationally by estimating the size of the smallest detail that has been determined, i.e. the resolution, or more generally, by comparing the structure with all known structural and biochemical data. A general computational approach is to use the Fourier shell correlation (FSC). This is particularly useful when little additional information is available or the resolution is modest (e.g. 15 Å). For this, the data are divided randomly into two sets and two independent reconstructions and their Fourier transforms are calculated. Equivalent shells in Fourier space from each structure are compared and the spacing at which the correlation becomes poor provides an estimate of the resolution achieved. For higher resolution structures (say better than 10 Å), in addition to FSC analysis, one can potentially trace out α-helices at $\sim$8 Å and β-sheets at 6 Å or better. Parts of the structure that may be known already from X-ray or NMR studies can be fitted and an estimate of resolution made.

The structure of several large complexes consisting of a number of proteins has been determined by electron microscopy and their interpretation requires establishing the boundaries and linkages between the various fragments, some of whose structures are known from X-ray diffraction or NMR studies. This is known as docking or fitting of the atomic models into the EM map and is often used to validate the quality of the microscopy. Examples are shown in Figures 2.7–2.9.

The structures obtained must be *presented* in an understandable and convincing manner, which again means representation of a three-dimensional structure as two-dimensional pictures. Surface shadowing software provides an iso-surface representation. These are essentially snapshots of the structure as if it were illuminated by a distant light source. The threshold used for the surface shading should be chosen carefully to represent the expected mass of the protein, or within one σ above the average density of the structure. Alternatively, the data may be presented in the form of a series of simple density sections, which additionally provide insight into details of the internal structure. Figure 2.7 shows surface shaded and density section representations of reovirus.

2.7 Software used in the analysis of electron micrographs

There are many software packages written for the analysis of EM data. They can be divided into those used for preprocessing images and those enabling a complete analysis to a three-dimensional reconstruction. The former group includes software for the selection of particles from micrographs, sometimes in an automated manner, and software to determine the defocus and state of stigmation in order to carry out a CTF correction. With the current explosion towards higher resolution, such programs are rapidly evolving to optimize effectiveness and speed. 'Ximdisp' for the manual selection of particles is currently the most user friendly and reliable program that can be run in stand-alone mode, whose output is fully compatible with many other packages. A useful program just for CTF determination is CTFFIND3, which automatically determines defocus and stigmation parameters.

Packages that allow a complete analysis of images are well established but each package was originally developed for a specific group of specimens. The MRC package was developed in the 1970s for the analysis of two-dimensional crystals and helices. With larger two-dimensional crystals it was possible to record electron diffraction patterns, which provided reliable amplitudes directly. In diffraction mode data are unaffected by specimen drift or the CTF of the microscope. Images of the crystals were used to derive phase information. To restore the complete three-dimensional distribution of amplitudes and phases in Fourier space, both electron diffraction and image data are required from several tilted specimens. This package permits the correction of lattice distortions (unbending), CTF correction, signal extraction from background noise, the merging of diffraction data recorded at different angles and calculation of the three-dimensional molecular density map.

Another group of packages analyses assemblies with helical symmetry. The first three-dimensional reconstruction method was developed for helical systems using a Fourier–Bessel approach. Today, the package Suprime is more commonly used and may be used to straighten filaments, analyse image spectra and perform three-dimensional analysis. However, some helical structures are very flexible and this results in changes in the helical parameters, i.e. disorder. Such difficulties are now being addressed using a single particle approach, where the helical assembly is cut into small elements that are analysed as single particles.

A number of packages relates to single particle analysis: these include the MRC package and its modified version Simplex, EM3DR, both of which are aimed at the analysis of structures with icosahedral or D5 symmetry. Many viruses display or enclose a capsid with icosahedral symmetry. Such structures may consist of one or several different proteins and vary in size from $\sim$200 to 2000 Å and are classified by a T number or a tiling approach. The icosahedral symmetry imposes a 60-fold degeneracy of information that permits the determination of a three-dimensional structure from just a few images. Most packages for the analysis of icosahedral structures use a Fourier–Bessel approach in contrast to other single particle packages, which use real space algorithms, primarily the back projection method. Packages such as IMAGIC-5, SPIDER, EMAN and Xmipp are used for analysis of single particle images with low or no symmetry. These packages utilise thousands of images of individual molecules. The software enables CTF correction, image alignment, orientation determination and the calculation of three-dimensional reconstructions.

Situs, URO, MOLREP, EMfit are packages that may be used for the *interpretation and objective fitting* of data. O and PyMOL are used mainly for manual fitting and inspection of data.

The software used to present results includes: surface representation (shading) Amira, Iris Explorer, PyMol, Chimera. Amira is an advanced software system for 3D visualization. Iris Explorer is powerful software that is particularly useful for surface shading and density interpretation. PyMOL is a molecular graphics system with an embedded script writer, and Python interpreter is, designed for real-time visualization and rapid generation of high-quality molecular graphics images and animations. It permits editing and fitting of PDB coordinates into three-dimensional maps. Chimera is an interactive molecular graphics program.

2.8   Examples

We have discussed the need for examining specimens close to their native state by cryo-EM and shown that high-resolution, but noisy, images result. After data analysis and image enhancement by computer processing, a three-dimensional reconstruction can be determined. We now present that state of the art with some of the highlights that have been achieved to date.

2.8.1   Two-dimensional crystals

These are assemblies of molecules that are organized crystallographically in two rather than three dimensions. They sometimes occur naturally, e.g. in bacterial membranes such as purple membrane or porin. For microscopy they are most easily interpreted when they are a just a single molecular layer thick. Considerable success has been achieved with membrane proteins. Their natural environment in the membrane permits detergent solubilization and subsequent crystallization in a lipid bilayer. Nonmembrane proteins may also form 2d crystals. Data are collected and analysed from samples tilted to ±60°. Symmetry of the molecule and crystallographic symmetry are used to minimize the effect of the uneven distribution of Fourier sections. The latest reconstruction of halorhodopsin has been obtained at 5 Å resolution. Y. Fujiyoshi and colleagues were able to determine a density map of bacteriorhodopsin at 3.5 Å resolution (Figure 2.5)

2.8.2   Helical structures

These may be considered as two-dimensional crystals that have been rolled up to form a cylinder with the unit cells matching at the line of connection, which may

(a)   (b)

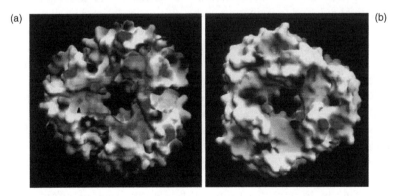

Figure 2.5   Illustration of a structure determined from analysis of 2D crystals at 3 Å resolution. Surface shaded views of (a) the cytoplasmic side and (b) the extracellular side of the bacteriorhodopsin trimer. Blue and red areas indicate positive and negative surface charges respectively. The arrow indicates the opening of the only proton channel that can be seen in this view. [Reproduced from Fujiyoshi, Y. (1999) *Faseb J.* **13** (suppl 2): S191–194]

(a)

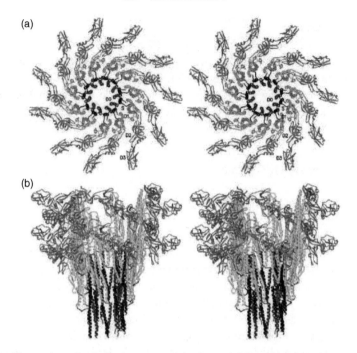

(b)

Figure 2.6 Illustration of a helical structure, the bacterial flagellar filament, determined at 4 Å resolution. Stereo views of the ribbon C_α backbone of the filament: (a) end-on view from the distal end of the filament with 11 subunits displayed; and (b) side view with three protofilaments removed for clarity. [Reproduced from Yonekura, K. *et al.* (2003) *Nature* **424**: 643–650]

be shifted by an integer number of unit cells. This is equivalent to describing a helix as being generated by the rotation of an element by a fixed angle α around an axis (the helical axis) in combination with a fixed translation in a direction parallel to that axis. This operation is repeated many times to generate the helix, in the manner of a crystallographic 'screw-symmetry' operator. Nature uses many helical structures, for example, cytoskeletol proteins such as collagen, actin and microtubules, phage particles whose tails have helical organization, bacterial flagella and amyloid fibrils which are associated with a range of human disorders including Alzheimer's disease. Near-atomic resolution has been achieved for the acetyl choline receptor with visualization of beta-sheets and for the R-type flagellar filament (Figure 2.6).

The analysis of *icosahedral structures* now routinely reaches 10 Å resolution and in the better cases is approaching the level achieved for 2D crystals and helical assemblies. This is, in part, due to the 60-fold redundancy of information, which means that each single particle image is equivalent to 60 particles with no symmetry. At the present time good results (6–8 Å) have been achieved for rice dwarf virus, hepatitis B virus and reovirus (Figure 2.7), enabling alpha helices to be traced.

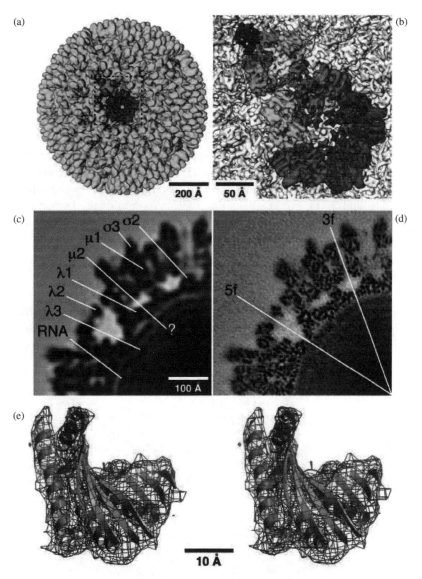

Figure 2.7 Example of single particle reconstruction with icosahedral symmetry at 7.6 Å resolution. Reovirus is shown in both surface-shaded and density section representation and known X-ray structures are docked into the virion. The structure is shown (a) surface-shaded at 20 Å resolution viewed down a 5-fold axis of symmetry. The $\lambda 2$ pentamer (blue) and a $\mu 1_3 \sigma 3_3$ heterohexamer ($\mu 1$, green; $\sigma 3$, red) are highlighted. (b) The outlined region of (a) at 7.6 Å resolution showing finer details with rod-like regions and the interlocking of the 5 $\lambda 2$ subunits. (c) The approximate location of the viral components in an equatorial density section 2.2 Å thick from the 20 Å rendering; (d) the same as (c) but from the 7.6 Å rendering with locations of 5-fold and 3-fold symmetry axes. Numerous punctate features arise from α-helices viewed end-on. (e) A stereo view of a small portion of the $\lambda 2$ X-ray ribbon structure fitted into the 7.6 Å EM map represented by the wire cage. α-Helices are shown in green and β-strands in yellow. [Reproduced from Zhang, X. et al. (2003) *Nat Struct Biol* **10**: 1011–1018]

2.8.3 Single particles with low symmetry

Protein complexes come in all sizes and symmetries. Spliceosomes and ribosomes, with molecular weights in the MDa range, are amongst the larger structures to have been analysed, the smaller are sub-complexes of spliceosomes, the transferrin complex and geminin with molecular weights in the 100 kDa range. Haemocyanines, helicases, portal proteins and heat shock proteins are in-between the two extremes: some of them have cyclical, others tetrameric or other symmetries (Figure 2.8). Each was analysed using the single particle approach. Orientational parameters were determined by projection matching or angular reconstitution. The structures determined for the ribosome, spliceosome and Gro-El (H. Saibil) are in the range 7–9 Å (Figures 2.10 and 2.11). For smaller molecules, cryo-data of the human transferrin receptor–transferrin complex (260 kDa) were analysed at ∼11 Å and human geminin (105 kDa) was analysed in negative stain at ∼17 Å resolution.

2.8.4 Cellular tomography

Recently a number of publications has demonstrated the feasibility of cryo-electron tomography (using an energy filter) to investigate larger, intact structures, such as the complete Herpes Simplex Virus (see Figure 2.12) and a small eukaryotic cell.

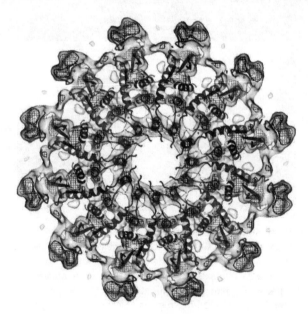

Figure 2.8 Cross-section of the connector of the SPP1 bacteriophage. The connector is a complex of the portal protein and two-head completion proteins. (courtesy of Dr. Orlova) The section shows the X-ray structure of the portal protein fitted into a cryo-EM map. The X-ray structure has been modified from 13-mer into 12-mer, because the connector within the bacteriophage capsid has 12-fold symmetry whereas the portal protein before incorporation into the procapsid has 13-fold symmetry. The excellent fit of density confirms the 10 Å resolution of the EM map

(a)

(b)

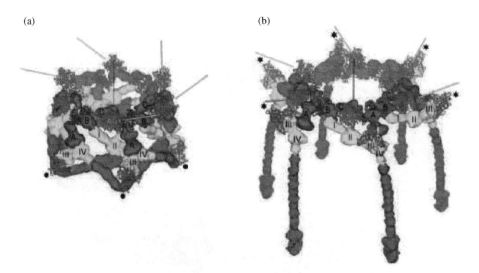

Figure 2.9 Illustration of conformational switching in the contraction of the tail of T4 phage when the baseplate switches from a hexagonal to a star conformation. (a) Structure of the periphery of the baseplate in the hexagonal conformation and (b) in the star conformation. Different proteins are identified by colour: gp7, red; gp8, blue; gp9, green; gp10, yellow; gp11, cyan; gp12, magenta. Combination of cryo-EM at 17 Å resolution and X-ray diffraction. [Reproduced from Leiman, P.G. et al. (2004) Cell **118**, 419–429]

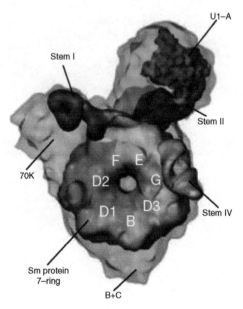

Figure 2.10 Three-dimensional structure of human spliceosomal U1 snRNP at 10 Å resolution. Tentative assignment of domains based on biochemical and structural information. The positions of the Sm proteins are indicated by a yellow ring. [Reproduced from Stark, H. et al. (2001) Nature **409**: 539–542]

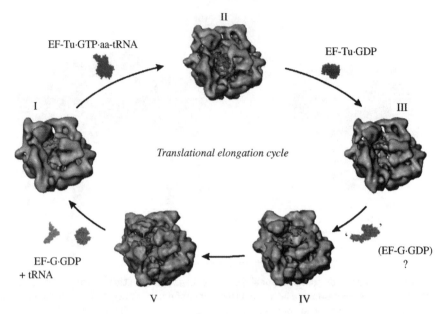

Figure 2.11 Illustration of single particle analysis of various states associated with the elongation cycle of the 70S *E. coli* ribosome (10–15 Å). EM data are shown in gold and other colours are docked X-ray structures of tRNA and various factors. [Reproduced from van Heel, M. *et al.* (2000) *Q. Rev. Biophys.* **33**: 307–369]

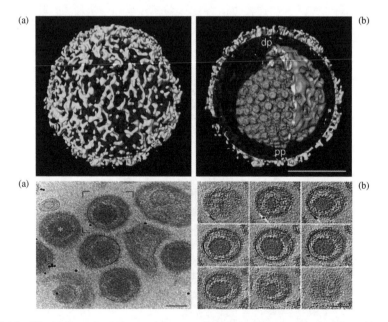

Figure 2.12 Illustration of a tomographic reconstruction of Herpes Simplex Virus. Upper panel: surface-shaded single virion tomogram; left, outer surface showing glycoprotein spikes (yellow) protruding through the membrane (blue) and, right, cutaway view showing the icosahedral capsid (light blue), the tegument (orange) and the envelope with spike proteins (blue and yellow). Lower panel: (a) untilted projected structure and (b) a series of slices through the particle boxed in (a). [Reproduced from Grunewald, K. *et al.* (2003) *Science* **302**: 1396–1398]

Although the resolution is limited to ~50 Å, this represents an exciting step towards systems biology with the visualization of smaller molecules such as the proteasome and Gro EL directly within the cell.

2.9 Conclusions

In the last decade the technique of Cryo-EM combined with image analysis has become a widespread and very powerful tool with which to examine biological structures close to their native state, approaching atomic resolution. Furthermore it enables molecules to be studied in a variety of different functional states using tiny amounts of material. In larger complexes linkages and active sites maybe precisely located in conjunction with ancillary X-ray or NMR data.

The ability to examine whole cells that is emerging with tomographic analysis is bringing the dream of understanding the dynamics of cells in terms of the function and localisation of proteins, their complexes and their relationship to organelles a step closer. Proteomics and systems biology are here with electron microscopy.

Acknowledgement

We thank Dr. Sarah Daniell for help with the bibliography and figures.

Further reading

Steven A.C. (1997). Overview of Macromolecular electron microscopy: an essential tool in protein structural analysis. *Current Protocols in Protein Science*. 17.2.1–17.2.29.

Baker T.S., Olson N.H., and Fuller S.D. (1999). Adding the third dimension to virus life cycles: three-dimensional reconstruction of icosahedral viruses from cryo-electron micrographs. *Microbiology and Molecular Biology Reviews*. 862–922.

Hayat, M.A. (1986). Basic techniques for transmission electron microscopy. Academic Press, N.Y.

Hawkes, P.W. (1972). Electron Optics and Electron Microscopy, Taylor and Francis: London.

Hawkes, P.W. (2004). Advances in Imaging and Electron Physics. Academic Press: London.

Herman, G.T. (1980). Image Reconstruction from Projections – the Fundamentals of Computerized. Tomography. Academic Press: London.

Henderson, R. (1995). The potential and limitations of neutrons, electrons and X-rays for atomic resolution microscopy of unstained biological molecules. *Q. Rev. Biophys*. 28, 171–193.

Dubochet, J., Adrian, M., Chang, J.-J., Homo, J.-C., Lepault, J., McDowall, A.W., and Schultz, P. (1988). Cryo-electron microscopy of vitrified specimens. *Quat. Rev. Biophys*. 21, 129–228

Fernandez-Moran, H. (1985). Advances in Electronics and Electron Physics, Vol. 16, pp. 167–223. Academic Press: London.

Frank, J. (1992). Electron tomography: three-dimensional imaging with transmission electron microscope. Plenum Press. New-York.

Frank, J. (1996). Three-dimensional electron microscopy of macromolecular assemblies. Academic Press. San Diego.

Frank, J. (2002). Single particle imaging of macromolecules by cryo-electron microscopy. *Annu. Rev. Biomol. Struct.* **31**, 303–319.

Moody, M.F. (1990). Electron microscopy of biological macromolecules. In Biophysical Electron Microscopy: Basic concepts and modern techniques (P.W. Hawkes and U.Valdre, eds.) Academic Press, New-York.

Crowther, R.A. (1971). Procedures for three-dimensional reconstruction of spherical viruses by Fourier synthesis from electron micrographs. *Phil. Trans. R. Soc. Lond. B.* 261, 221–230.

Crowther, R.A. & Amos, L. A. (1971). Harmonic analysis of electron images with rotational symmetry. *J. Molec. Biol.* **60**, 123–130.

Fuller, S.D., Butcher, S.J., Cheng, R.H., and Baker T.S. (1996). Three-dimensional reconstruction of icosahedral particles – The uncommon line. *J. Structural Biol.* **116**, 48–55.

Radermacher, M. (1988). Three-dimensional reconstruction of single particles from random and non-random tilt series. *J. Elec. Microsc. Tech.* **9**, 354–394.

Parsons, D.F. (1970). Some Biological Techniques in Electron Microscopy, Parsons, D.F., ed., pp. 1–68. Academic Press: London.

Reimer, L. (1997). Transmission Electron Microscopy – Physics Of Image Formation and Microanalysis. Springer: Berlin.

van Heel M, Gowen B, Matadeen R, Orlova EV, Finn R, Pape T, Cohen D, Stark H, Schmidt R, Schatz M, Patwardhan A. (2000). Single-particle electron cryo-microscopy: towards atomic resolution. *Q. Rev. Biophys.* **33**(4), 307–369.

3

Atomic force microscopy: applications in biology

James Moody and **Stephanie Allen**

Laboratory of Biophysics and Surface Analysis, School of Pharmacy, The University of Nottingham, Nottingham, UK

3.1 A brief history of microscopy

The invention of the optical microscope in the early 1600s magnified and focussed light from objects, for the first time allowing people to see the wonders of life on the microscale. Although fascinating, some scientists could see no practical relevance, dismissing the tool as a whimsical novelty. Clearly this viewpoint could not last for long and by 1665 Robert Hooke published his groundbreaking book *Micrographia*. Hooke's compound design for the microscope, combining two or more lenses, changed little over the coming centuries.

The resolving power of optical microscopes, referring to the ability to successfully distinguish two points as being separate, is ultimately limited by the wavelength of the visible light being used. However, in 1923 Louis de Broglie, a French mathematical physicist, proposed that matter could be thought of as having a wavelength, which in the nonrelativistic limit is inversely proportional to its momentum. Thus it was postulated that electrons could be used to bombard a sample's surface and, as a consequence of their small wavelength, potentially provide a much greater level of image resolution. This was realized in the late 1930s with the development of the first working scanning electron microscope (SEM).

Whilst electron microscopy is still widely used today for the routine analysis of biomolecular samples, it has limitations. First, nonconducting samples must be made conductive, usually achieved by applying a thin layer of gold to the surface. Second,

Chemical Biology   Edited by Banafshé Larijani, Colin. A. Rosser and Rudiger Woscholski
© 2006 John Wiley & Sons, Ltd

electrons are easily scattered or absorbed by air molecules, so specimens must be viewed in a vacuum. This requires removing or fixing any water in a sample, as the water may vaporize in the vacuum. This is fine for inorganic samples, but biological specimens generally contain water and are nonconductive. These requirements for sample preparation mean the SEM is therefore not best suited for the analysis of biological samples in their native state. Even despite relatively recent developments to overcome such issues, including the development of cryogenic electron microscopy (cryo-EM) or environmental scanning electron microscopy (ESEM), these limitations ultimately resulted in the need for an alternative method of sample analysis.

3.2   The scanning probe microscope revolution

The atomic force microscope (AFM) belongs to a class of instruments known as scanning probe microscopes (SPM), that obtain information using a sharp probe or tip to directly 'feel' the surface of a given sample. Although the 1980s take the credit for developing the high-resolution scanning probe microscopes that we see today, the principles behind their development date back as far as the 1920s, and in particular to the stylus profiler (developed by Schmalz in 1929).

3.2.1   The stylus profiler

This instrument utilized an optical lever arm mounted on the end of a cantilever to monitor the movement of a sharp probe over the surface of sample. By recording the motion of the stylus it was possible to generate images with up to 1000 times magnification. The main drawback of this method was the possible bending of the probe when it encountered large surface features. In 1971, Russell Young demonstrated a form of noncontact stylus profiling by exploiting the fact that the electron field emission current between a sharp metal probe and an electrically conductive sample is dependent on the distance separating them. For his profiler, which he named the topographiner, the probe was mounted directly on a piezoelectric ceramic used to raster-scan the surface in the x–y plane. To overcome the obstacle of surface features, an electronic feedback loop was employed to adjust the vertical position of the probe. By monitoring the probe in the z-direction whilst scanning across the surface in the x–y plane, a detailed three-dimensional image could be constructed.

3.2.2   The scanning tunnelling microscope

In 1981, Binnig and Rohrer applied the principles employed in the topographiner to develop the first SPM, the scanning tunnelling microscope (STM). By monitoring the quantum tunnelling of electrons from the sample surface to the probe, detailed information about the sample could be gathered in a similar fashion to the topographiner. However, quantum tunnelling is far more sensitive than electron field emission and requires the probe to scan in very close proximity to the sample surface, greatly increasing the resolution of the image compared with that obtained by the topographiner. Using this technique, Binnig and Rohrer were able to image individual silicon atoms on a surface.

The STM is still widely used today, in particular for the analysis of conducting or semi-conducting substrates. The resultant STM images are most often related to sample topography, but it should be noted that the technique in fact directly measures the localized charge densities of a surface. Although the STM has also been successfully employed for biomolecular imaging in a small handful of studies, the requirement for the sample to be conductive has meant that its use in this area is somewhat limited.

3.2.3 The atomic force microscope

A major advancement came in 1986, with the development of the AFM, which was based on a principle similar to that of the stylus profiler. By using an ultrasmall probe mounted on a cantilever, high-resolution images could be produced without the need for a conducting surface. Of all the descendents of the STM, it is the AFM that is still the most widely used, as it allows the determination of surface properties in a variety of media, including air and liquid. Operating in liquid is especially useful for observing biological specimens, as buffers can be used *in vitro* to closely mimic native conditions. This allows the observation of biological processes which would not be possible with other SPMs.

3.3 The workings of an AFM instrument

The AFM is fundamentally different from an optical microscope in its approach to imaging. Instead of looking at the sample with light, the AFM essentially 'feels' the sample with a sharp probe. In an optical microscope, light is used to instantaneously illuminate the whole of a sample's surface, allowing a complete image to be constructed immediately. The AFM, however, has only one probe and must gather information about the sample by touching each part sequentially. This is done in an ordered fashion by raster-scanning the imaging probe over the surface (or the surface under the probe). Thus, although the AFM is slower at imaging than an optical microscope, in addition to surface topography it can also potentially extract more information regarding the sample material and/or interfacial properties.

3.3.1 The imaging probe

The central part of the AFM is the imaging probe. Its size and sharpness are critical to the level of attainable image resolution, and to limit imaging artefacts the probe should be sharper than the features being imaged. The AFM probe is a sharp micro-fabricated spike (the radius of curvature of the apex is approximately 10–20 nm) mounted on the end of a flexible cantilever. The cantilever is a spring which deflects in response to forces acting between the probe and sample, as the probe follows the contours of the surface. For biomolecular imaging low spring constant cantilevers (e.g. 0.01–0.1 N/m) are typically employed, allowing the user to accurately minimize the level of applied force and thus avoid potential irreversible damage of the sample. Cantilevers and probes are nearly always made from either silicon or silicon nitride, due to their robustness and ease of fabrication.

AFM probes fall into two categories: namely, high and low aspect ratio. The aspect ratio refers to the length of the probe divided by its apex width. The choice of probe depends on the sample being imaged; for a rough sample, a high aspect ratio probe is required, whereas for a relatively flat sample, a low aspect ratio probe can be used (Figure 3.1).

The shape of the cantilever on which the probe is mounted also varies, with two basic geometries dominating. The triangular-shaped cantilever is designed to reduce torsional rotation whilst scanning, avoiding possible twisting when encountering objects on the surface. The second common geometry is a rectangular beam. This type of cantilever allows torsional rotation and can be used to measure the frictional properties of a sample.

3.3.2 The piezoelectric scanner

Modern AFMs use piezoelectric transducers to either raster-scan the imaging probe across the stationary sample, or the sample with respect to a stationary probe. Certain crystals exhibit a property known as the piezoelectric effect. When compressed or stretched, a piezoelectric crystal will build up alternate charges on opposite faces.

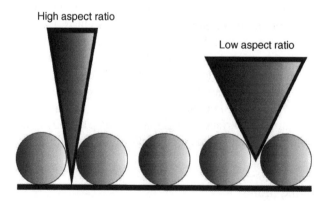

Figure 3.1 A high aspect ratio probe is capable of imaging rough samples, as it is able to image between the surface features (shaded spheres). A low aspect ratio probe is not able to do this as well and is thus more suited to flatter surfaces

Alternatively, if a potential difference is applied across opposite faces, the crystal will expand or contract. By using this property, a piezoelectric scanner can accurately control movement in the x-, y- and z-directions, with sub-nanometre resolution. The response time for the piezoscanner can become a limiting factor when observing fast real-time processes.

3.3.3 The deflection detection system

One of the simplest and most efficient methods for amplifying the tiny deflections of the cantilever during imaging relies on a laser-beam being reflected from the back of the cantilever to a position-sensitive photodiode (Figure 3.2). Any change in the bend of the cantilever will move the position of the reflected laser-spot centred on the photodiode. The change in the bend of the cantilever corresponds to the displacement

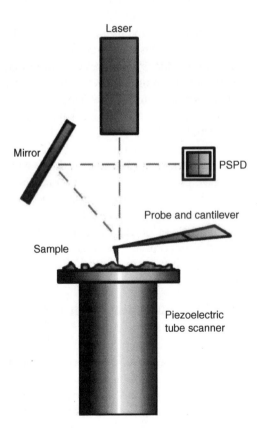

Figure 3.2   A schematic diagram of a typical AFM instrumental set-up. A laser is aligned to reflect off the back of the cantilever onto a position-sensitive photodiode (PSPD), which allows the accurate tracking of the small cantilever deflections which result as the probe is scanned over the sample surface (or, as in the case of this instrument, the sample is scanned beneath a stationary probe)

of the laser beam from its initial spot on the photodiode. The photodiode is divided into four quarters, upon which the laser beam is centred before imaging can begin. By comparing the intensity of light falling on the top half of the photodiode with the bottom half, any change to the bend of the cantilever can be monitored. By comparing the left quadrants of the photodiode with the right quadrants, twisting of the cantilever can be monitored (e.g. for the imaging of surface frictional properties).

3.3.4   The electronic feedback system

While scanning the surface of a sample, the probe will encounter features due to the presented topology, resulting in small deflections of the cantilever. In the most commonly employed 'constant force' imaging method, the instrument aims to keep the level of cantilever deflection, and thus the applied force, constant throughout the imaging process. This is achieved by combining information from the position-sensitive photodiode with a feedback control loop. In this constant force mode, upon deflection of the cantilever away from the user-defined level of deflection (known as the 'set point'), the feedback loop adjusts the z-position of sample or probe, in order to restore the level of deflection to the original value. In this mode, the resultant three-dimensional image is generated by plotting the feedback loop adjustments made in the z-direction at each point scanned in the x–y plane.

It should be noted, however, that there will be a finite time associated with this process, i.e. there will be a small lag between the initial deflection signal and the point at which it is restored to its original value due to the response time of feedback loop. For optimal operation, the AFM should return the cantilever to the desired set-point as quickly as possible to allow the probe to accurately track the surface, otherwise the information used to build up the image would convey a different topology to that being analysed. To overcome this, the instrumental user is able to manually adjust the response time of the feedback loop during imaging, via the feedback loops gains (commonly referred to as the proportional, integral and derivative gains).

3.4   Imaging biological molecules with force

At the fundamental basis of atomic force microscopy are the forces which occur between the sample and the imaging probe. The Lennard-Jones potential (Figure 3.3) describes the interaction between two uncharged molecules or atoms as they approach one another. When two uncharged particles approach one another, the electron clouds undergo a disruption. This distortion of the uniform electron clouds means the particles no longer have a homogeneous distribution of charge. The resulting induced dipole moment causes the two particles to become mildly attracted to one another (i.e. by van der Waals attractive forces). As the two particles become closer together, their electron orbits eventually begin to overlap and a repulsive force begins to dominate. The repulsive force arises because the bound electrons

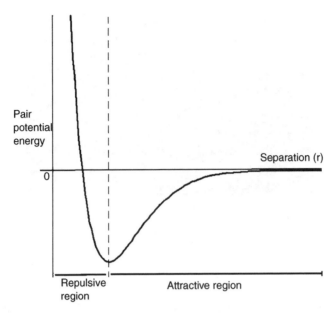

Figure 3.3 The Lennard–Jones pair potential is a good approximation for the potential energy between two approaching particles

surrounding atoms or molecules are forbidden from entering the same quantum state as the bound electrons from other atoms or molecules.

It is within these two force regimes that the AFM essentially operates. By working at small probe–surface separations (i.e. when the probe is in direct physical contact with the surface), repulsive forces arise between the atoms in the probe and sample surface. This method of imaging is known as contact mode. Alternatively it is also possible to image with the probe located further from the surface where attractive probe–sample forces dominate. In this region the specimen can be imaged without the probe ever coming into contact with the surface, and hence this imaging method is known as non-contact mode.

3.4.1 Contact mode

In contact mode, the AFM probe is brought into direct physical contact with the sample surface. The probe is traced along the surface, and the cantilever deflects in response to changes in the repulsive probe-sample forces which occur due the sample's topology. The amount of force applied to the sample during imaging is chosen by the user through the instrument software (i.e. through adjustment of the set-point). For biomolecular imaging, the level must be optimized for each sample, as too high a force can result in irreversible damage. The level of force applied will also influence the quality of the obtained image. For example, if the biomolecular species of interest is only weakly attached to the underlying surface, during contact mode imaging it may be dragged along or 'swept' by the probe. Not only does this process lead to the production of a low quality image, but it can also result in the

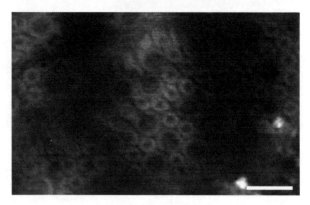

Figure 3.4   A high resolution contact mode AFM image of a native photosynthetic membrane (128 nm) of *Rsp. photometricum* at submolecular resolution (scale bar: 20 nm; z-scale: 6 nm) (From Scheuring *et al*. The EMBO J. (2004) 23, 4127, Reproduced by permission of the Nature Publishing Group)

contamination of the probe with sample material, which can subsequently lead to a reduction in the sharpness of the probe and/or worse, introduce artefactual features in the obtained image (see Sections 3.5.2 and 3.5.3).

The high repulsive and lateral forces associated with contact mode imaging thus tend to preclude its use for the imaging of isolated biological species, in particular if they are only loosely attached to the surface on which they are deposited. However, it should be noted that in this repulsive force regime the probe is actually most sensitive to small changes in force, and thus offers the best potential for high-resolution imaging. To take advantage of this for biomolecular imaging it is therefore necessary to considerably reduce the imaging forces used (e.g. by imaging in liquid, see Section 3.4.4) and prepare samples so that they are more resistant to deformation and/or potential damage by the imaging probe. It is this approach that has been employed to obtain some of the highest resolution images of biomolecules that have been reported in the literature. In such studies the authors have imaged to sub-nanometre resolution highly ordered two-dimensional arrays or highly compact films of proteins such as bacteriorhodopsin and light harvesting complexes (see Figure 3.4). The observed high level of image resolution is attributed to (i) the close-packing of the proteins, which makes them more resistant to deformation by the probe and (ii) the manipulation of the electrolyte concentration of the buffered aqueous imaging environment, which enables the authors to effectively balance attractive and repulsive probe–sample forces to a point at which resolution is optimized.

3.4.2   Oscillating cantilever imaging modes

To circumnavigate the problem of dragging poorly immobilized specimens across the substrate and/or the potential damage of delicate samples, alternative imaging modes were subsequently developed. The first of these was termed 'non-contact' mode. In this mode the cantilever is oscillated above the surface at its resonant frequency, with low amplitude. During imaging, the oscillating probe scans close to the surface, but is never allowed to actually touch the sample. Any increase in the long range attractive forces which act at these probe–sample separations results in a slight dampening of the cantilever oscillations. As this relationship is proportional, in non-contact mode it is the change in cantilever oscillation amplitude that is typically monitored throughout imaging, rather than cantilever deflection. The disadvantage of this, however, is that the relationship is only weakly dependent on probe-sample separation, and thus image resolution tends to be lower than that of contact-mode.

A second closely related imaging mode termed intermittent contact or 'tapping' mode, was thus also developed. Here the cantilever is oscillated at a higher amplitude so that the probe taps the surface at the lowest point of its oscillation period. As a consequence this mode is able to take advantage of the higher resolution associated with the repulsive forces on probe-sample contact, but is also able to dramatically reduce the lateral forces imposed on weakly immobilized features during imaging. Consequently, for the imaging of biological molecules and other delicate samples, tapping mode has become an invaluable tool.

A useful by-product of oscillating cantilever imaging modes arises from the ability to also monitor the difference in phase between the oscillations of the driving signal and the cantilever. The resultant phase-images are typically simultaneously obtained with tapping mode images, and can provide additional information about the material properties of the sample surface. For example, when tapping a hard surface, the difference between the phase driving signal and the phase of the cantilever is minimal, as little energy is lost when hitting the hard surface. However, if the properties of the surface suddenly change to become more adhesive or viscoelastic, more energy will be absorbed when the probe interacts with the surface, causing the cantilever oscillations to move further out of phase. This mode of imaging becomes particularly useful when the topography of the sample is relatively flat, for example when imaging polymer blends (see Figure 3.5).

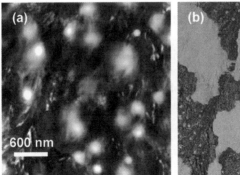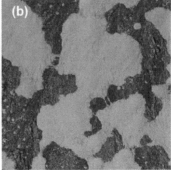

Figure 3.5 Simultaneously acquired (a) topography and (b) phase images of a poly(sebacic anhydride) (PSA)–poly(lactic acid) (PLA) 80:20 polymer blend obtained using tapping mode (in air). Although some detail can be observed in the topography image, the differentiation between the crystalline PSA and amorphous PLA regions on the sample is significantly more pronounced in the phase image (image provided courtesy of Professor X. Chen, University of Nottingham)

3.4.3 Imaging in liquid

Although images of biomolecular samples can be obtained in air, this approach is somewhat limited due to the influence of capillary forces that arise between the probe and sample. The probe and features on the surface act as nucleation sites for the condensation of water vapour present in the air, and at small probe–sample separations small capillaries form between them, which manifest as large forces during imaging (of greater than 10^{-7}N). Such forces are eliminated when samples are imaged in liquid environments. Imaging in air is of course, also not generally the best option for biological specimens. Under native conditions, biomolecules exist in a liquid environment; their stability arises from this medium and the processes in which they are involved occur within this surrounding. By allowing the observation of biomolecules in aqueous buffers that can be used to mimic native conditions, the AFM has set itself apart from other techniques for the investigation of biological specimens. To image biomolecules under these conditions, a liquid cell is used to contain the sample solution (e.g. the biomolecule suspended in aqueous buffer) and to maintain a clear optical path for the laser beam to track changes in the cantilever deflection.

3.5 Factors influencing image quality

Although the AFM has become an invaluable tool for the analysis of biological specimens, no technique is free from limitations. However, by understanding the limitations of the technique being applied, it is possible to recognize and avoid problems that may arise. The following text describes the main factors that can influence the quality of biomolecular AFM images and how they may be recognized and/or overcome.

3.5.1 Sample preparation and immobilization

Sample preparation will ultimately determine the quality of the images observed. For example, poorly prepared samples may show nothing but debris due to contamination of the surface. Cleanliness during sample preparation and imaging is therefore essential, and all aqueous solutions should be filtered prior to use (and at most used 1 month after preparation).

Immobilization of the biological molecule to the underlying substrate is also critical when imaging to prevent its displacement/removal by the probe. To this end a range of approaches have been developed including physical adsorption, electrostatic and covalent attachment methods. The method employed is dependent on the imaging environment, the substrate employed and the properties of the biomolecule of interest. The most commonly employed substrates include mica, gold and silicon/ silica, as these are flat and have well defined chemistries which facilitate bio-molecular deposition or attachment. For example, for the imaging of DNA, two immobilization methods have proved to be particularly useful. One method utilizes an unmodified mica surface (negatively charged) with a buffer containing divalent metal ions such as Ni^{2+} or Mg^{2+}, or mica presoaked/pretreated with solutions of such ions. Here the divalent ion acts to bridge the negatively charged DNA backbone to the mica substrate. The other approach uses mica or alternative flat substrates (e.g. silicon or gold) onto which positive charges have been introduced, for example through chemical functionalization with aminosilane or alkylthiol reagents.

It should also be noted that the method of surface attachment may also have an influence the properties of the biomolecule of interest. For example, while covalent attachment may facilitate the structural investigation of isolated biomolecular species, such a strong attachment may be disadvantageous if the user wishes to subsequently visualize any dynamic processes of this molecule, as they may be hindered/prevented. For the observation of such dynamic processes the use of immobilization methods such as weak electrostatic interaction is therefore more usual, and in this type of experiment thorough investigation and optimization of the immobilization process is absolutely essential.

3.5.2 Tip convolution/broadening

When imaging a relatively large surface feature, the AFM probe can be thought of as infinitely sharp. This is, however, clearly not the case as the probe has real physical dimensions. In this particular example, the size of the feature is simply so much bigger than the probe that it is essentially 'dimensionless'. However, when imaging features that are similar in size to the probe, the resulting image obtained is actually a convolution of the physical size of the feature and the probe. The extreme case of tip convolution is when the sample features are sharper than the imaging probe, so that the surface feature now effectively images the AFM probe (rather than the more normal opposite situation). This situation typically manifests in the resultant image as many repeats of exactly same feature (i.e. the image of the probe apex) in the location of such sharp surface features, and is thus relatively easy to identify.

The apex of an AFM probe typically has a radius of curvature of around 10–20 nm. When imaging an object whose width is hundreds of nanometres, the width of the feature observed in the image will not be much different to its actual width. However, many biological samples such as isolated proteins, actin filaments or dsDNA helices have dimensions similar to the radius of curvature of the probe. When analysed using the AFM, the resulting images of such molecules therefore typically show broader dimensions than predicted by other analytical techniques. This is because imaging of the feature actually commences upon contact with the side of the imaging probe, rather than the apex itself (see Figure 3.6). As the lateral position is measured from the apex of the probe, analysis of the data will indicate the width of the feature to be greater than it actually is.

The recorded width of the dsDNA helix, for example, when analysed using AFM is typically 12–15 nm, whereas a width of 2 nm is predicted from the crystal structure. For specimens that are approximately circular in cross-section, it is sometimes

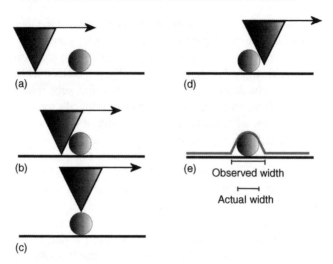

Figure 3.6 A schematic explaining the origin of probe–sample convolution. As the probe approaches the object to be imaged, (a) and (b), the side of the probe makes contact first, causing a premature rising of the cantilever. A similar phenomenon occurs as the probe passes over the imaged feature, (c) and (d). The path of the probe apex is shown in (e), and the difference between observed width of the feature and its actual width is shown

therefore more appropriate to estimate their width by measuring the feature height. However, height measurements can also be problematic as many biological molecules suffer from compression and/or deformation by the force exerted by the imaging probe. Such compression may also increase the actual width of the sample, and thus further complicate the interpretation of AFM image data.

As tip–sample convolution occurs due to the finite size of the probe, it is obvious that the effect would be reduced by further miniaturizing the probe's dimensions. With some success, recent attempts to achieve this goal have utilized single-walled carbon nanotubes attached to the apexes of regular AFM probes. Commercially available sharpened silicon oxide probes can also help improve the attainable level of resolution.

3.5.3   Double tipping

An effect commonly referred to as 'double tipping' occurs when the apex of the sharp probe is fractured and/or is contaminated. For 'double-tipping' the probe effectively has two apexes, although 'multi-tipping' can also occur if more asperities are present. When using a tip like this to image a surface, both probes will also encounter the imaged surface feature, causing the feedback loop to adjust the deflection of the cantilever twice. This leads to an image providing two copies of the encountered features, with a separation equal to the distance between the two tips. This problem is easily recognized owing to the repetition of the surface features in the resulting image (see Figure 3.7).

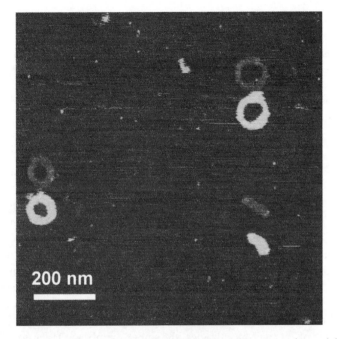

Figure 3.7 A tapping mode AFM image of plasmid DNA–polymer complexes (obtained using tapping mode in aqueous buffer) showing 'double tipping'. The repeating features allow this type of imaging artefact to be easily recognized (image provided courtesy of J.E. Ellis, University of Nottingham)

3.5.4 Sample roughness

The physical dimensions of the probe can also limit the level of sample roughness that can be imaged. A typical probe is only around 3 μm long. This allows only surfaces where features are less than 3 μm tall/deep to be imaged; for taller/deeper features, the underside of cantilever will end up touching parts of the sample. However, there are certain ways to overcome this limitation, e.g. when trying to image a 4 μm diameter spherical cell with a probe around 3 μm long. By packing many of the same cells on the surface, i.e. creating an effective monolayer, the probe would only have to fall and rise by a maximum of 2 μm, allowing the top hemispheres to be imaged.

3.5.5 Temperature variation and vibration isolation

Owing to its very nature the AFM can be susceptible to changes in temperature and external vibrations. Providing these variables are realized and steps are taken to minimize their effects, a user should still be able to obtain high-quality images. Air conditioning can help to minimize the effects of thermal fluctuations by stabilizing the temperature of the room within which the AFM is operated.

Environmental chambers in which the AFM can be housed are also available, if external temperature variation becomes problematic. To minimize the problem of external vibrations inherent in any building, the AFM instrument can be placed on pneumatic vibration isolation tables or other cheaper alternatives, such as platforms suspended on bungee cords. Such approaches are commonly employed in SPM laboratories and reduce a large fraction of unwanted vibrations that may interfere with imaging.

3.6   Biological applications of AFM and recent developments

Since its invention, the AFM has been used to image many biological systems including nucleic acids, membrane proteins and more complex protein–nucleic acid assemblies. Cell biologists have also harnessed the AFM's unique capabilities to study the dynamic behaviour of living and fixed cells ranging from red and white blood cells to living renal epithelial cells. For example, *in vivo* plasma membrane turnover of migrating epithelial cells has been imaged in real time. Other approaches have also been developed that take advantage of the AFM's ability to interact with/ manipulate the imaging substrate. For example, by increasing the scanning force on isolated patches of rat liver membranes, the upper membrane layer could be stripped from the extracellular surface, revealing hexagonal arrays of gap junctions that could be subsequently imaged. These studies, performed *in situ* in phosphate buffered saline (PBS), were able to provide images with a resolution of less than 3 nm.

Experiments such as these have demonstrated the versatility of the AFM's application to provide information on biomolecular structure and biological processes. In recent years, the ability of the AFM to visualize dynamic biological processes and to record the forces involved in biomolecular interactions have created particular interest and are thus discussed in more detail below.

3.6.1   Imaging dynamic processes

By recording a series of images over the same region of a sample, it is possible to monitor biological processes as a function of time and build up animations of real-time processes. This approach has, for example, enabled the study of processes which include protein fibrillization, the transcription of DNA by RNA polymerase, the translocation of molecules through membrane pores/channels and the enzymatic degradation of potential gene-delivery polymer–DNA complexes.

A major problem when imaging such dynamic processes, however, is the limiting factor of scan speed. Conventional AFMs take around 1–2 min to produce an image. Dynamic processes occurring on time scales faster than this cannot therefore be successfully observed. However, recently developed smaller cantilevers allow faster imaging as their higher resonant frequencies and reduced dimensions permit higher scanning speeds. Recent instrumental advancements in scanning speed have also enabled images to be captured at rates comparable to video-rate. Although at present most of these images have been obtained in air, through further instrumental development it will be possible to routinely follow fast real-time processes as they occur in conditions similar to their natural environment.

3.6.2 Measuring biomolecular forces

In addition to imaging, the AFM can be used to measure the forces acting between the probe and the substrate. In aqueous environments the AFM is routinely able to measure forces as small as 10–20 pN. For the investigation of biomolecular samples, the small probe–sample contact areas and force-measurement range provided by the AFM make it exquisitely suited for the study of single biomolecular interactions or properties.

Force measurements between individual biological receptor–ligand complexes can be obtained by functionalizing the AFM probe with the receptor and the opposite substrate with the complementary ligand. The surfaces are brought into contact allowing the formation of the receptor–ligand complex. The probe is then separated from the surface and the force required to disengage the complex is measured. The main challenge in such experiments is to achieve the measurement of a single molecular interaction, rather than the interaction of several biomolecular pairs. A number of approaches have been used to achieve this requirement including the reduction of binding site availability via the addition of free ligand to the experimental environment and dilution of the surface density of the interaction molecules. Statistical analysis of the frequency at which biomolecular interaction events are recorded can also provide a method of estimating the likelihood of single molecule interaction measurement.

The streptavidin–biotin complex was the first receptor–ligand interaction to be investigated using such an approach, due to its high affinity and well characterized nature. The initially reported values for the streptavidin–biotin unbinding forces were, however, found to vary considerably (from 83 to 410 pN). Further studies have since demonstrated that this variability is due to the sensitivity of this type of measurement to the rate of bond rupture (or more correctly the rate at which the bond is loaded). This relationship has given rise the related experimental approach, termed dynamic force spectroscopy, in which the rupture forces of receptor–ligand complexes are typically recorded as a function of measurement speed. The resultant dynamic force spectra (plots of the most commonly observed force vs the logarithm of the loading rate) can be used to reveal information on the dynamic response of the bond to force, and the locations and heights of energy barriers along the force-induced dissociation pathway. Consequently, dynamic force spectra for a range of biomolecular systems have been recorded including a variety of cell-adhesion molecules, and for DNA and RNA oligonucleotides.

In another related approach, often referred to as single molecule force spectroscopy (SMFS), force measurements have also been used to determine the intramolecular forces within a range of biopolymeric molecules, including proteins, polysaccharides and nucleic acids. In such measurements, a molecule becomes attached to the probe surface during probe-sample contact (typically through physisorption) and then it is mechanically extended as the probe and sample are separated. The response of this molecule during this process results in deflection of the cantilever, and allows one to monitor, for example, the force required to induce conformational changes as a function of distance. This approach has been used to study conformational changes in polysaccharides such as dextran, and in nucleic acids and their complexes with other molecules, but has found particular application in the protein folding/unfolding field.

The first protein unfolding experiments were carried out on the large muscle protein titin. Titin, a multidomain protein composed of tandem repeat units of fibronectin type III and immunoglobulin, forms filaments approximately 1 μm in length and is a major constituent of the sarcomere in vertebrate striated muscle. When AFM was used to perform force–extension measurements on individual titin molecules, periodic 'saw-tooth' rupture events were observed, with a periodicity of 25–28 nm. Each 'saw-tooth' event was attributed to the unfolding of the individual protein domains. Since these early experiments, improvements upon this experiment have been made, principally to overcome the heterogeneity of the domains within the titin molecule. The approach now adopted employs engineered polypeptides comprising identical protein domains, which permits more defined investigations of mechanically induced protein unfolding mechanisms (see Figure 3.8). Combined with a dynamic force spectroscopy approach and the ability to construct single or multiple amino-acid mutations of the protein under investigation, it is therefore now possible to employ protein SMFS measurements to develop sophisticated models

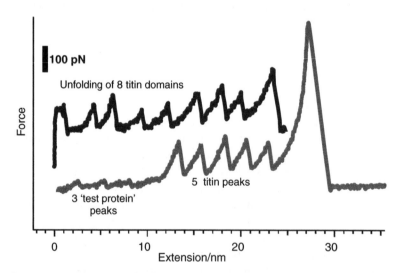

Figure 3.8 Representative 'saw-tooth' protein unfolding traces for protein constructs containing either eight copies of a titin immunoglobulin domain (upper trace) or five copies of a titin immunoglobulin domain, and three copies of a 'test-protein' (lower trace). The peak of each 'tooth' in these traces corresponds to the unfolding of one protein domain. The presence of the titin domains in the construct unfolded in the lower trace provides an internal standard for comparison to the unfolding events of the other 'test' protein domains [force curve provided courtesy of Dr W. Zhang, Dr P.M. Williams (University of Nottingham) and Dr. J. Clarke (University of Cambridge)]

describing unfolding energy landscapes. This may be particularly relevant for proteins and other biological molecules that have evolved to be responsive to force, and hence may provide new insights into the origin of their behaviour.

3.7 Conclusions and future directions

Since its invention, the AFM has evolved into a powerful tool for the investigation of the structure, mechanical properties and interactions of biological molecules. A particular area where the approach still holds considerable future potential is for the investigation of membrane proteins, which currently present many nontrivial challenges for more conventional experimental approaches. Indeed the particular value of using both high-resolution imaging and SMFS measurements for the study of such proteins has been recently demonstrated in the literature. Through future investigations of such molecules, together with the information derived from experiments on the other biomolecules involved in cellular interactions and function, it will therefore be possible to obtain a new class of fundamental information on the structure and interactions that are vital in maintaining the biological environment.

Further reading

J. Clarke and P.M. Williams (2005). Unfolding induced by mechanical force. In *Protein Folding Handbook. Part I*, Edited by Buchner, J. and Kiefhaber, T. Wiley-VCH: New York.

H. Clausen-Schaumann, M. Seitz, R. Krautbauer and H.E. Gaub (2000). Force spectroscopy with single bio-molecules. *Curr. Opin. Chem. Biol.* **4**: 524–530.

P.L.T.M. Frederix, T. Akiyama, U. Staufer, Ch. Gerber, D. Fotiadis, D.J. Müller and A. Engel (2003). Atomic force bio-analytics. *Curr. Opin. Chem. Biol.* **7**: 641–647.

H.G. Hansma, K. Kasuya and E. Oroudjev (2004). Atomic force microscopy imaging and pulling of nucleic acids. *Curr. Opin. Struct. Biol.* **14**: 380–385.

R. Krautbauer, L.H. Pope, T.E. Schrader, S. Allen, H.E. Gaub (2002). Discriminating drug–DNA binding modes by single molecule force spectroscopy. *FEBS Letts* **510**: 154–158.

D.J. Müller, D. Fotiadis, S. Scheuring, S. A. Müller and A. Engel (1999). Electrostatically balanced subnanometer imaging of biological specimens by atomic force microscope. *Biophys. J.* **76**: 1101–1111.

S. Scheuring, J. Seguin, S. Marco, D.Lévy, C. Breyton, B. Robert and J.-L. Rigaud (2003). AFM characterization of tilt and intrinsic flexibility of rhodobacter sphaeroides light harvesting complex 2 (LH2). *J. Mol. Biol.* **325**: 569–580.

4

Differential scanning calorimetry in the study of lipid structures

Félix M. Goñi and **Alicia Alonso**

Unidad de Biofísica (Centro Mixto CSIC-UPV/EHU) and Departamento de Bioquímica, Universidad del País Vasco, PO Box 644, 48080 Bilbao, Spain

4.1   Introduction

Differential scanning calorimetry (DSC) is a technique that allows the measurement of the minute exchanges of heat (in the microcalorie range) that accompany certain physical processes in biomolecules, such as aggregation/dissociation, conformational changes or ligand binding. DSC does not perturb the system with chemical probes, is often nondestructive, and can nowadays be performed with very small amounts of material, i.e. tens of micrograms. Typical applications of DSC include protein studies such as thermal stability and thermal denaturation, but in this chapter we shall focus our attention on the use of DSC in the field of lipids. We shall start by reviewing briefly some aspects of lipid structure and properties, particularly in relation to their role in membranes. Then a few basic points of thermodynamics and calorimetry will be put into the context of membrane biophysics.

4.2   Membranes, lipids and lipid phases

Cell membranes are composed of a double layer (or bilayer) of amphipathic lipids with which a variety of proteins are associated. Amphipathic lipids are thus called because they contain both a hydrophobic and a hydrophilic moiety. Lipids in membrane bilayers are oriented in such a way that their hydrophobic (lipophilic)

Chemical Biology   Edited by Banafshé Larijani, Colin. A. Rosser and Rudiger Woscholski
© 2006 John Wiley & Sons, Ltd

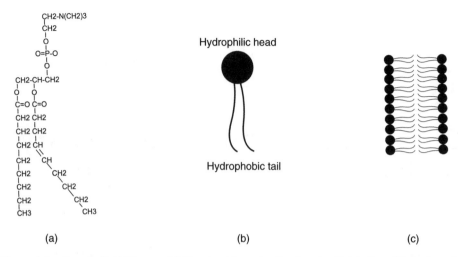

(a)   (b)   (c)

Figure 4.1 Phospholipid bilayers. (a) Structural formula of a phosphatidylcholine. (b) Outline of a phospholipid, indicating the polar head and the hydrophobic tails. (c) Outline of a phospholipid bilayer, in which the polar headgroups are in contact with the aqueous medium while the hydrophobic tails interact with each other

regions are in contact with each other, while their hydrophilic portions interact with the aqueous environment (Figure 4.1).

Membrane lipids are *soluble* only in organic solvents, such as chloroform or methanol. Only in these are the individual molecules of lipids solvated and dispersed. Membrane lipids are not soluble in water, because of their hydrophobic components. However, their hydrophilic moieties interact with water molecules. The result is a series of complex structures that are spontaneously formed when membrane lipids come into contact with water. These structures are interpreted in terms of lipid phases.

In general, a phase is defined as any homogeneous part of a system that is physically distinct being separated from other parts of the system by a definite boundary, in thermodynamic equilibrium, e.g. water molecules at 1 atm, between 0 and 100 °C are in the liquid phase; at higher temperatures water is in the vapour phase.

In two-component lipid–water systems, and under the biological conditions of excess water, i.e. H_2O being over roughly 50 per cent of the mass, the following phases have been observed: monolayers, lamellar, micellar, hexagonal and cubic. All membrane lipids, even at very low concentrations, form *monolayers* in excess water. In monolayers lipids orient themselves at the air–water interface, with the hydrophobic part oriented towards the air, and the hydrophilic portion in contact with the water molecules. At higher concentrations the interface is saturated with lipids. At this point the sterols (such as cholesterol) begin precipitating out of the system. Other membrane lipids (phospholipids, sphingolipids) form one of the other phases, according to their molecular structures. Most membrane lipids, when in pure form, adopt the *lamellar* phase, and the same occurs with virtually all lipid mixtures

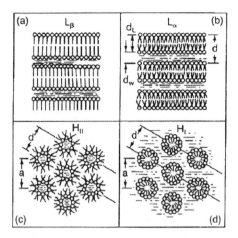

Figure 4.2 Lipid phases. (a) Gel (L_β) lamellar phase. (b) Liquid-crystalline (L_α) or fluid lamellar phase. (c) Inverted hexagonal (H_{II}) phase. (d) Hexagonal phase. [Reproduced from R.B. Gennis (1989) *Biomembranes: Molecular Structure and Function*. Springer: New York, (p. 41), with permission]

extracted from cell membranes [Figure 4.2(a,b)]. In fact, membrane bilayers at equilibrium have the structure of lamellar phases. (Note, however, that phases as thermodynamic concepts are abstract, limitless entities, while membrane bilayers are finite and real.)

Some membrane lipids, in excess water, give rise to nonlamellar structures. Thus lysophosphatidylcholine (a phosphatidylcholine lacking the fatty acid at the *sn*-2 position) forms *micelles*, in common with many synthetic detergents and with bile salts, the natural detergents structurally related to sterols. Micelles are spherical or ellipsoidal objects, in which the constituent lipids have their hydrophobic moieties forming a lipophilic core, and their polar parts interacting with water at the surface. Some gangliosides (glycosphingolipids with a complex sugar moiety) also give rise to micelles when dispersed in water. A few glycerophospholipids, e.g. phosphatidylethanolamine, and some lipid mixtures, become organized in a tubular fashion in water. The transversal section of the tubes has a hexagonal form, with the inner part of the tubes being filled with water and the outer part staying occupied by the hydrocarbon chains. This is a *hexagonal* phase, specifically the one so-called, for technical reasons, 'inverted hexagonal' or H_{II} phase [Figure 4.2(c)]. Finally, a very few lipids, or lipid mixtures, when swollen in water give rise to one of the inverted cubic phases.

The lamellar phase is particularly important because, as mentioned above, it represents more or less the structure of cell membranes under equilibrium conditions. It should be added now that the name 'lamellar phase' is not fully accurate because there are at least five lamellar phases, namely the subgel or crystalline phase, the gel or solid phase, the rippled phase, the fluid or liquid-crystalline phase, and the liquid-ordered phase. Lipids in the subgel and in the *gel lamellar phase* (sometimes denoted

L_β) are highly ordered and virtually immobile [Figure 4.2(a)]. Only some lipids with highly saturated hydrocarbon chains, mostly of synthetic origin, can adopt a gel phase above 0 °C. A few phospholipids (saturated phosphatidylcholines and phosphatidyl-glycerols) can exist, at temperatures intermediate between the gel and fluid phases, in the so-called *rippled phase* ($P_{\beta'}$), characterized by the ripples seen by electron microscopy in freeze-fractured surfaces of this phase. In the *liquid-crystalline*, or fluid, or liquid disordered phase (L_α) [Figure 4.2(b)], the lipids diffuse freely in the plane of the membrane, and the flexible hydrocarbon chains exhibit a large degree of motion. Most glycero- and sphingolipids in cell membranes are probably in a physical state akin to a liquid-crystalline phase. The *liquid-ordered phase* (L_o) cannot be attained by any pure lipid, but exists instead in mixtures of glycerolipids or sphingolipids with sterols, provided the hydrocarbon chains of the former are sufficiently saturated. In the L_o phase the lipids diffuse laterally in the plane of the membrane, but their hydrocarbon chains are restricted in their rotational motion, i.e. they are relatively ordered, hence the name of this phase.

Phase changes, or *phase transitions*, as they are usually called, may occur when the defining parameters (composition, pressure, temperature) are changed. For instance, when the temperature of water at 1 atm and 90 °C increases to 110 °C, there is a phase transition from liquid to vapour phase. Transitions induced by providing heat to or removing it from a system are called *thermotropic*. Phase transitions may be caused through changes in pressure (barotropic transitions), or proportion of solvent (lyotropic transitions). DSC is, of course, used in the detection of thermotropic transitions. The best characterized thermotropic transitions in membranes and membrane lipid–water systems are the interconversions between the various lamellar phases (gel to fluid, gel to rippled, rippled to fluid) and the lamellar to inverted hexagonal transition. These have very important biological implications, as will be discussed below (Sections 3 and 4).

4.3 Heat exchanges and calorimetry

4.3.1 Heat and related entities

Heat is a form of energy that can be intuitively associated to molecular motion. Transferring heat to a system is tantamount to increasing the motion of its constituent molecules. Heat has been traditionally measured in calories (cal), one calorie being the amount of heat required to increase the temperature* of 1 g of liquid water by 1 °C at 1 atm. However, in order to stress the interconvertibility of all forms of energy, it is better to express heat in joules (J), the international unit for energy, 1 J being approximately equivalent to 0.24 cal.

---

*Note that, in spite of the obvious relation between heat and temperature, they are very different entities. Heat is an extensive entity (i.e. it varies with the amount of matter) while temperature is an intensive entity.

As mentioned above, thermotropic phase transitions may be brought about at constant pressure by adding or removing heat. (In biology, atmospheric pressure is considered constant for all practical purposes.) This is the case when ice is heated up to give liquid water, or when a lipid bilayer in the L_β phase is heated up to give an L_α phase. In thermodynamics, the amount of heat gained or lost by a system at constant pressure is called *enthalpy* (H). We can write:

$$\Delta H = \Delta q_p$$

where q_p is the heat *absorbed* by the system at constant pressure. Consequently ΔH is always positive for endothermic processes (i.e. processes in which heat is absorbed). Enthalpy has units of energy.

A related useful parameter is C_p, or *heat capacity* at constant pressure. Heat capacity refers to the amount of heat required by a system in order to have its temperature increased by a given amount, at constant pressure. Its units are energy temperature^{-1}.

Within a given phase, C_p can be thought of as a constant for most practical purposes. For instance for water at 1 atm, between 0 and 100 °C, C_p has a fairly constant value of ~4.18 J/°C. However, when going from liquid water to vapour water, i.e. at a thermotropic phase transition, C_p increases enormously, because at or near 100 °C heat is supplied to the system with little change in temperature, being used instead to bring about the phase transition. The opposite, i.e. a decrease in C_p, is seen when a transition is induced by cooling down a system, e.g. going from water vapour to liquid water in the previous example. Thus if we could measure C_p for a given system at different temperatures, thermotropic phase transitions would be observable as abrupt changes in C_p at given temperatures, the *transition temperatures*.

Box 1 Analysing DSC thermograms

The quantitative analysis of DSC data of lipid samples is rather straightforward. Three main parameters are derived from the thermograms, namely transition temperature, transition width and transition enthalpy (see Figure in this box). All three are routinely given by the software that accompanies any contemporary instrument.

The *transition temperature* should be measured at the *onset* of the transition (T_{on}, also called T_c). However, for broad transitions, the temperature is measured at the midpoint of the transition, i.e. in the middle of the peak (T_m).

The *transition width* is measured, in °C, at midheight of the endotherm ($\Delta T_{1/2}$). This gives a measure of the sharpness (or cooperativity) of the transition.

The *transition enthalpy* ΔH is obtained, by definition, through integration of the C_p curve (i.e. the thermogram) between the onset and completion temperatures.

The transition enthalpy obtained directly from the thermogram is called 'calorimetric enthalpy' (ΔH_{cal}). In addition, a related parameter, the 'van't Hoff enthalpy' (ΔH_{VH}) can be approximately obtained from the relationship:

$$\Delta H_{VH} \approx 4RT_m^2/\Delta T_{1/2}$$

The ratio $\Delta H_{\mathrm{VH}}/\Delta H_{\mathrm{cal}}$ gives the *cooperative unit* (in molecules) that measures the degree of intermolecular cooperation between phospholipid molecules. For maximum cooperativity (first-order phase transition) this ratio should approach infinity.

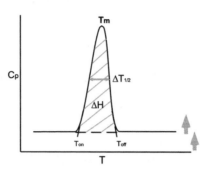

Box 1   Useful thermodynamic parameters derived from a DSC thermogram

4.3.2   Differential scanning calorimetry

Differential scanning calorimetry is a technique that measures C_{p} as a function of temperature. Thus it is ideally suited to detect thermotropic phase transitions in lipid–water systems (as indeed in many other biological and nonbiological samples). In lipid and membrane studies, high-sensitivity DSC instruments are used that allow liquid samples, down to 100 μM in lipid (detection of certain transitions with small ΔH, such as the L_{α}–H_{II} transition, see below, may require higher lipid concentrations) and with volumes of ~0.5 ml, to be measured. A typical DSC instrument and an outline of its components are shown in Figure 4.3. In a typical instrument two identical tantalum cells exist, one containing only buffer. The system is kept under positive pressure (e.g. N_2 gas) to avoid bubble formation during heating. Both cells are heated up at a constant rate through the main heaters, whilst monitoring the temperature differences between the sample (S) and reference (R) cells. These differences are compensated for by the feedback heaters (see Figure 4.3). The extra heat energy uptake required by the sample cell to maintain the same temperature as the reference cell corresponds to the excess heat capacity occurring during an endothermic phase transition in the lipid. (The opposite would happen with an exothermic process, i.e. less heat would be required by the sample than by the reference sample to keep a steady change in temperature).

This basic instrumentation can be complemented in various ways. For instance, sudden pressure changes may be applied to the system, thus obtaining additional thermodynamic information from the so-called 'pressure perturbation calorimetry'. In this method the heat consumed or released after a pressure jump is measured, and from this volume changes associated with phase transitions may be derived. Another interesting modification is

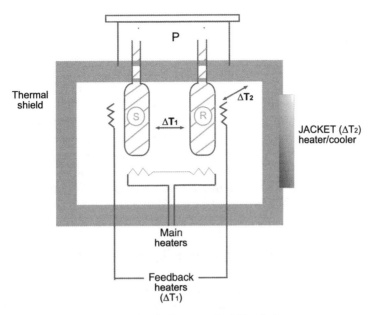

Figure 4.3   Sketch diagram of a DSC calorimeter

called 'capillary DSC', which reduces considerably sample volume (down to 0.1 ml) and permits automated analysis of multiple samples ('high throughput analysis').

The standard DSC experiment with lipid systems requires degassing the samples and instauring an extra pressure of 1–2 atm N_2. The cells and the surrounding adiabatic thermal shield are then equilibrated at least 5 °C below the desired start temperature of the experiment. Heating starts at a pre-established rate (e.g. 1 °C/min) in a region where no transition is expected, so that a baseline is obtained. At a temperature above the phase transition the whole system is cooled down, re-equilibrated and re-scanned. Three heating scans are routinely recorded; sometimes the first scan differs from the other two, owing to insufficient equilibration of the sample, in which case the first scan is discarded.

4.4   Phase transitions in pure lipid–water systems

Most glycerolipids and sphingolipids in aqueous dispersions form closed vesicles, limited by lipids in the lamellar (bilayer) disposition. Depending on the lipid structure, different thermotropic transitions may be observed, of which the following are the most common.

4.4.1   Tilted gel ($L_{\beta'}$) to rippled ($P_{\beta'}$) phase transition (the 'pre-transition')

Some saturated phospholipids (phosphatidylcholines and phosphatidylglycerols) in the gel phase have their hydrocarbon chains somewhat tilted with respect to the normal of the bilayer, instead of orienting them parallel to the normal, as most lipids in the gel phase do. The tilted gel phase is denoted $L_{\beta'}$. $L_{\beta'}$ phases can undergo a thermotropic transition in which the hydrocarbon chains become oriented parallel to the bilayer normal. As a result a global rearrangement of the bilayer ensues with the appearance of the rippled structure that is characteristic of the $P_{\beta'}$ phase. Rippled structures are probably the result of interactions between parallel bilayers in multilamellar structures. For dipalmitoylphosphatidylcholine (DPPC), i.e. a phosphatidylcholine containing two esterified palmitic (hexadecanoic) acids, perhaps the best studied phospholipid of all, the $L_{\beta'}$–$P_{\beta'}$ transition, sometimes called the pre-transition, occurs at 35 °C, and the associated change in enthalpy is ~6 kJ/mol. Similar thermodynamic parameters for other phospholipids can be found in Table 4.1.

4.4.2   Rippled ($P_{\beta'}$) to liquid-crystalline (L_{α}) phase transition

The rippled phase may, upon further absorption of heat, become a liquid-crystalline phase. This phase transition is called the phospholipid 'main transition' because of its large change in C_P, as compared with the other transitions in the same system. It occurs at $T_m = 41$ °C in DPPC, and ΔH is in this case ~36 kJ/mol. See Table 4.1 for the T_m and ΔH values of other lipids. The liquid-crystalline, or fluid, or liquid-disordered phase is thought to represent the state in which most membrane lipids exist.

Table 4.1   Transition temperatures of phospholipids in excess water

Lipid	Transition	$T(°C)$	ΔH (kJ/mol)
diC12 PC	Gel–fluid	7.0	16
diC14 PC (DMPC)	Gel–fluid	24	26
diC16 PC (DPPC)	Gel–fluid	41.5	36.5
diC18 PC (DSPC)	Gel–fluid	55	46
diC20 PC	Gel–fluid	67	50
diC18:1 PC (DOPC)	Gel–fluid	−16	32
diC18:2 PC	Gel–fluid	−53	—
diC16 PC (DPPC)	$L_{c'}$–$L_{\beta'}$	18.5	17.6
diC16 PC (DPPC)	$L_{\beta'}$–$P_{\beta'}$	34	5.9
diC12 PE	L_{β}–L_{α}	29	17
diC14 PE (DMPE)	L_{β}–L_{α}	46	25
diC16 PE (DPPE)	L_{β}–L_{α}	63.5	34.5
diC18 PE (DSPE)	L_{β}–L_{α}	74.5	44
diC18:1(*cis*) PE (DEPE)	L_{β}–L_{α}	38	26.5
diC18:1(*cis*) PE (DEPE)	L_{α}–H_{II}	65	1.7
diC14 PG (DMPG)	$P_{\beta'}$–L_{α}	23	25.1
diC16 PG (DPPG)	$P_{\beta'}$–L_{α}	40	31.5
diC16 PS (DPPS)	Gel–fluid	54	37

PC, phosphatidylcholine; PE, phosphatidylethanolamine; PG, phosphatidylglycerol; PS, phosphatidylserine.

4.4.3 Gel (L$_\beta$) to liquid-crystalline (L$_\alpha$) phase transition

Saturated glycero- and sphingophospholipids other than phosphatidylcholines and phosphatidylglycerols, that cannot form tilted L$_{\beta'}$ phases, form instead the gel L$_\beta$ phase. In these cases, the gel-to-fluid L$_\beta$-to-L$_\alpha$ thermotropic transition (without a rippled intermediate) has very similar properties to that of the main transition in, e.g. DPPC (see Table 4.1).

The transition temperatures (T_m) of both the P$_{\beta'}$–L$_\alpha$ and L$_\beta$–L$_\alpha$ transitions are very sensitive to the nature of the fatty acyl hydrocarbon chains, the temperatures decreasing with chain length, and, very markedly, with the presence of *cis*-double bonds in the fatty acyl residues. This is shown, for example, by the decrease in T_m in the progression: distearoyl (C18:0), dioleoyl (C18:1), dilinoleoyl (C18:2) phosphatidylcholines (Table 4.1).

4.4.4 Lamellar (L$_\alpha$) to inverted hexagonal (H$_{II}$) phase transition

Some phospholipids, typically phosphatidylethanolamines (PE), can undergo under certain conditions a thermotropic L$_\alpha$–H$_{II}$ transition that can be conveniently monitored by DSC.

Lamellar-hexagonal transitions occur, for a given system, at higher temperatures (T_h) than gel–fluid transitions. For example, for dielaidoylphosphatidylethanolamine (DEPE), $T_m \approx 38\ °C$, and $T_h \approx 65\ °C$. The enthalpy change, ΔH, associated to the L$_\alpha$–H$_{II}$ transition is smaller than that occurring during the L$_\beta$–L$_\alpha$ transition. ΔH values for the transitions in DEPE are 27 kJ/mol (L$_\beta$–L$_\alpha$) and 2 kJ/mol (L$_\alpha$–H$_{II}$). Unsaturation has a profound effect on the L$_\alpha$–H$_{II}$ transition of PE. For instance, distearoyl PE remains in the bilayer phase to over 100 °C, while dioleoyl PE undergoes the L$_\alpha$–H$_{II}$ transition at ~10 °C. As will be shown below through specific examples, C_p changes that accompany the L$_\alpha$–H$_{II}$ transition are very sensitive to the presence of other components, lipids or proteins, in the bilayer.

4.5 Selected examples of transitions in lipid mixtures

When two lipids are co-dispersed in excess water, the situation becomes more complicated. A simple tool used in the analysis of phases and phase transitions in multicomponent systems is the *phase diagram*. For condensed systems (containing only solid and liquid phases, no vapour phases) the effects of pressure can be neglected, and phase diagrams become simpler. A two-component condensed system may be depicted by a rectangular diagram, in which the mole fraction of each component is given in the abscissa, and temperature in the ordinate. Each phase occupies a given area in the rectangle. Phase boundaries are represented by lines. Similarly, a three-component condensed system at a given temperature may be represented by a triangular diagram.

Phase diagrams of two- and multicomponent lipid mixtures are complicated by at least three different factors:

Box 2   Lipid–DNA complexes for gene delivery

Delivery of recombinant DNA to eukaryotic cells is not a trivial procedure. A method that has been used to help in the transformation includes presenting DNA in the form of complexes (*lipoplexes*) with cationic lipids, e.g. DOTAP *N*-([1-(2,3-dioleoyloxy)propyl])-*N,N,N*-trimethylammonium chloride. Lipoplexes form spontaneously when cationic liposomes and DNA come into contact. One important parameter of lipoplexes, namely hydration, has been measured by D. Hirsch-Lerner and Y. Barenholz using DSC. In this case free water content is computed from the large ice–water transition endotherm at ca. 0 °C. The endotherm for the lipoplex is smaller (as low as 50 per cent in some cases) than the sum of the endotherms for each of the components, showing that dehydration occurs during lipoplex formation. Dehydration happens to be a prerequisite for the intimate contact between cationic lipids and DNA, and is probably instrumental in facilitating DNA transport across the cell membrane.

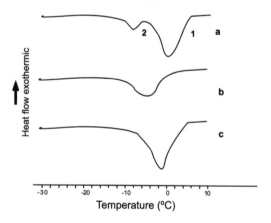

Box 2 Thermograms of cationic liposomes, DNA, and lipoplexes. (a) DOTAP–DOPE (mole ratio 1:1) in H_2O. Peak 1: gel–fluid transition of the lipid. Peak 2: melting of free water. (b) Plasmid DNA (0.95 mg) in excess water (1.31 mg). (c) Lipoplex with a DNA–lipid charge ratio of 1.5. The above thermograms indicate an amount of bound water of (a) 14 H_2O molecules per lipid molecule, (b) 12 H_2O molecules per DNA phosphate group, (c) 12.5 H_2O molecules per (DNA phosphate group + lipid molecule). [Adapted from D. Hirsch-Lerner and Y. Barenholz (1999) *Biochim. Biophys. Acta* **1461**: 47–57, with permission.]

(1) *Nonideal miscibility* – in ideal mixtures, the components do not interact with each other, thus the properties of the mixture depend linearly on the corresponding property of each component, and on the composition of the mixture. However, in practice, two-component lipid mixtures behave nonideally, and this gives rise to anomalies. For example the phosphatidylcholines dimyristoyl PC (DMPC, C14 saturated fatty acids), dipalmitoyl PC (DPPC, C16 saturated fatty acids) and distearoyl PC (DSPC, C18 saturated fatty acids) have their main

transition temperatures at $\sim$23, 41 and 55 °C, respectively. DMPC and DPPC mix almost ideally, thus an equimolar mixture of DMPC and DPPC has a transition temperature at $(23 + 41)/2 = 32$ °C. However DMPC and DSPC do not mix ideally, and in fact two different transition temperatures, ca. 31 and 44 °C, are seen in the DSC thermogram.

(2) *Phase coexistence* – based in part on the phenomenon of nonideal miscibility, but also for other reasons, real phase boundaries in the phase diagram are often not lines, but strips or areas in which two or more different phases coexist. Lipid transitions are not sharp, but have instead a finite width, i.e. they extend over a certain temperature range. Specific examples will also be presented below.

(3) *Complex formation* – up until now we have assumed that the various lipids did not interact with each other (ideal behaviour) or at least they did not interact stoichiometrically. However, stoichiometric complex formation is sometimes a possibility that complicates the phase diagrams, as will also be shown below.

4.5.1 Phospholipid–cholesterol mixtures

Cholesterol is well-known to be involved in the genesis of cardiovascular disease. Less well-known, but perhaps even more important, is its implication in healthy cellular events, namely the regulation of molecular order in cell membranes. This occurs mostly through interactions of cholesterol with phospholipid molecules in the membrane bilayers. DSC happens to be particularly useful in the study of those interactions.

It was mentioned above that DPPC was the best studied of phospholipids. Cholesterol is perhaps the best studied of all lipids. Thus it will be no surprise that DPPC–cholesterol is the binary lipid mixture that has been most exhaustively explored. It is indeed a very difficult system. Even mixing and swelling properly the lipids in water is far from trivial; remember that sterols by themselves cannot be dispersed in water (Section 4.2), although they can be co-dispersed with phospholipids, giving rise to bilayers. The DPPC–cholesterol system was first studied using DSC in the 1960s by Dennis Chapman. He made the fundamental observation that cholesterol had the effect of gradually widening and decreasing the main thermotropic transition of DPPC at 41 °C, to the point of totally abolishing it at equimolar DPPC: cholesterol ratios. A plethora of studies has followed to this day, using DSC together with many other techniques, and the behaviour of this deceptively simple mixture is not yet fully understood. All the possible complications mentioned above (nonideal behaviour, phase coexistence, complex formation), and perhaps several others, occur in this binary mixture. For instance, in the presence of increasing amounts of cholesterol, the transition temperature hardly decreases, in contrast to what would be observed if cholesterol formed an ideal mixture with DPPC in the L_α phase and did not dissolve in the DPPC $L_{\beta'}$ phase.

Perhaps the most detailed study of DPPC–cholesterol mixtures by DSC is the one performed by McMullen and McElhaney, whose phase diagram is reproduced in

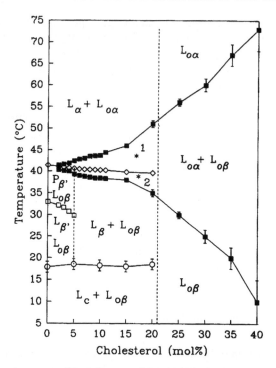

Figure 4.4   Temperature-composition diagram of DPPC-cholesterol in excess water. [Reproduced from T.P.W. McMullen and R.N. McElhaney (1995) *Biochim. Biophys. Acta* **1234**: 90–98, with permission]

Figure 4.4. In this phase diagram we find the $L_{\beta'}$, $P_{\beta'}$, L_{β} and L_{α} phases that have already been described, plus a few new ones, namely $L_{c'}$, $L_{o\alpha}$ and $L_{o\beta}$. $L_{c'}$ (denoted c in the figure) is the pure DPPC crystalline or subgel phase, that is only observed in 'annealed' (i.e. kept at 4 °C for several days) aqueous multilamellar dispersions of DPPC. A thermotropic transition, the 'subtransition', converts the $L_{c'}$ into the $L_{\beta'}$ phase at ca. 18 °C. $L_{o\alpha}$ and $L_{o\beta}$ are two different liquid ordered (L_o) regions, respectively called liquid-crystalline-like liquid ordered ($L_{o\alpha}$) and gel-like liquid ordered ($L_{o\beta}$). The latter two phases differ in the orientational order of the hydrocarbon chains, and in the relative position of the cholesterol molecule in the host PC bilayer.

The diagram in Figure 4.4 can be divided into two regions, respectively 'outside' and 'inside' the main endotherm, marked by '■'. The regions outside the endotherm, i.e. above and below the lines connecting the black squares, will be described first. Incorporation of cholesterol at temperatures below the subtransition causes the appearance of the DPPC–cholesterol $L_{o\beta}$ phase, which coexists with extended arrays of pure DPPC in the $L_{c'}$ phase. Increasing amounts of cholesterol gradually decrease the enthalpy of the subtransition, indicating that the size and/or number of $L_{c'}$ domains is decreasing until they disappear at approximately 20 mol per cent cholesterol (cholesterol mole fraction 0.2). Similarly at temperatures between the subtransition and the pre-transition (i.e. 18 and 34 °C), pure DPPC exists in the $L_{\beta'}$

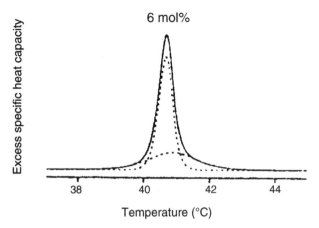

Figure 4.5 Sharp and broad components in a DSC thermogram. The sample corresponds to DPPC–cholesterol (94:6 mole ratio) in excess water. [Reproduced from McMullen and McElhaney (1995), with permission]

phase, but addition of cholesterol causes the formation of the $L_{o\beta}$ phase. $L_{\beta'}$ and $L_{o\beta}$ coexist up to 5 per cent cholesterol, when the pretransition disappears, and an untilted L_β phase is formed that coexists with $L_{o\beta}$ in the 5–20 per cent cholesterol concentration range. The rippled $P_{\beta'}$ also gives rise in part to $L_{o\beta}$ until, above 5 per cent cholesterol, $P_{\beta'} + L_{o\beta}$ are substituted by $L_\beta + L_{o\beta}$. Finally, at temperatures above 41 °C, the pure DPPC L_α phase coexists, in the presence of cholesterol, with $L_{o\alpha}$, up to 20 per cent cholesterol. At higher cholesterol concentrations, the liquid ordered $L_{o\beta}$ and $L_{o\alpha}$ phases exist on their own, respectively at low and high temperatures.

The main endotherm consists of two superimposed components, a sharp and a broad one (Figure 4.5). The sharp transition is usually attributed to the melting of cholesterol-poor domains, and the broad one is attributed to the melting of cholesterol-rich domains. The boundaries of the broad transition are marked with '■' in the phase diagram (Figure 4.4), and the sharp transition temperatures are indicated by '◇'. 'Inside' the broad transition, and at cholesterol concentrations up to 20 mol per cent, these are two subregions, above and below the sharp transition. Three phases coexist in each of these sub-regions, namely L_β, $L_{o\alpha}$ and $L_{o\beta}$ (below) and L_α, $L_{o\alpha}$ and $L_{o\beta}$ (above). As the level of cholesterol increases beyond 20 mol per cent, only the ordered phases persist. At and above 50 mol per cent cholesterol (outside the graph shown in Figure 4.4) pure cholesterol domains form, and a cooperative lipid transition is no longer observable.

Similar DSC studies have been carried out with a related binary mixture, namely sphingomyelin/cholesterol. Different sphingomyelins (SM) vary in melting temperature, but all of them appear to have a gel–fluid transition. N-palmitoyl SM, the main component of egg SM, has a transition at 41 °C. Its mixtures with cholesterol behave similarly to the DPPC/cholesterol system, in that: (a) cholesterol causes the endotherm to broaden and its enthalpy to decrease without marked changes in the transition temperature; (b) the transition is abolished at or above 50 mol per cent cholesterol, and (c) mixtures containing up to 20 mol per cent sterol can be deconvolved into a sharp and a broad component, that are interpreted again as arising from the transition of cholesterol-poor and –rich domains, respectively.

Two major controversies in the interpretation of data from phospholipid–cholesterol mixtures are (i) whether SM does or does not interact with cholesterol *preferentially* to PC, and (ii) the nature of the bilayer components giving rise to the broad component of the main calorimetric transition. Based on DSC, but also on spectroscopic studies, some authors find that cholesterol may interact more strongly with SM than with PC molecules of comparable hydrocarbon-chain length and structure, and therefore preferentially stabilize the gel over the fluid phase. Other authors, using similar techniques, interpret their results as suggesting no preferential affinity of cholesterol for SM over PC in miscible SM/PC bilayers. The matter is still unsettled. The other on-going discussion is the origin of the broad heat absorption that may be due to the melting of cholesterol-rich domains, as was originally proposed, or to the thermal dissociation of *condensed complexes* of phospholipid and cholesterol, as suggested more recently by McConnell. Note that the condensed complexes need not form a separate phase; the complexes can exist within a homogeneous phase. It is difficult at present to distinguish between the two possibilities, and yet the answer may be important for understanding a number of phenomena in cell membranes.

A final point concerning DSC studies of phospholipid–cholesterol mixtures is the finding of cholesterol crystallites, that melt in the 80–95 °C range, in SM–cholesterol mixtures with cholesterol concentrations of 60–80 mol per cent. This is a relevant observation because the lens of the eye in mammals contains SM, dihydrosphingomyelin and cholesterol in high concentrations. Cholesterol crystallites may play a functional role in vision.

4.5.2   Lamellar-to-inverted hexagonal transitions

The lamellar phase represents the structure of cell membrane lipids under steady-state conditions. However in certain circumstances, particularly in membrane fusion events (e.g. in egg fertilization, or cell infection by some viruses), membrane lipids abandon transiently the lamellar disposition, adopting nonlamellar architectures, of which the best known is the so-called 'inverted hexagonal', or H_{II}, phase. Nonlamellar structures are at the origin of the 'lipid stalk', a structural intermediate that connects two bilayers in the membrane fusion process. Only certain lipids, or lipid mixtures, can undergo the L_{α}–H_{II} thermotropic transition, and the latter can be detected by DSC. H_{II}, like other nonlamellar phases, has received particular attention lately because of its possible implication in important phenomena such as cell membrane fusion, or protein insertion into membranes. High-sensitivity DSC instruments allow the detection of L_{α}–H_{II} transitions with phospholipid suspensions of concentration 5 mM or even less.

L_{α}–H_{II} transitions have been observed mainly in PE or related phospholipids. Egg PE at pH 5.0 exhibits one such transition at ~30 °C that is observable by DSC. DEPE at neutral pH has a L_{α}–H_{II} transition at ~65 °C, and *N*-monomethyl dioleoylphosphatidylethanolamine (DOPE-Me) has the corresponding transition at ~63 °C.

The lamellar-hexagonal transition of PE is very sensitive to the presence of other lipids in the bilayer. Even small proportions (<5 mol per cent) of alkanes of six to 20 carbon atoms dramatically broaden the transition endotherm, and lower the T_h transition temperature, i.e. facilitate the probability of the transition at a given temperature. Addition of 5 mol per cent dodecane to DOPE-Me decreases the L_α–H_{II} transition temperature from ~63 to ~0 °C. This is believed to occur because alkanes (like other apolar molecules) tend to occupy the 'voids' that would exist between the lipid tubes in the inverted hexagonal phase, thus effectively stabilizing the H_{II} phase. In general, the potency of the alkanes in reducing T_h increases with chain length. In fact, other nonpolar molecules also facilitate the L_α–H_{II} transition by lowering the transition temperature. This is the case of alkanols (C6–C12), e.g. 20 mol per cent dodecanol is required to decrease the T_h of egg PE by ~20 °C. Fatty acids are also effective in facilitating the L_α–H_{II} transition, particularly at slightly acidic pH, when these molecules are mostly protonated. Unsaturation of the fatty acyl chains greatly increases their H_{II} phase-forming ability, perhaps because *cis*-double bonds introduce 'kinks' in the chain, so that the molecule can more easily accommodate to fill in the voids in the H_{II} phase.

Perhaps more relevant from the physiological point of view is the facilitating effect of diacylglycerols (DAG) and ceramides on the L_α–H_{II} transition (Figure 4.6). Both types of lipids broaden markedly the transition endotherm, and decrease the transition temperature, DAG being more potent than ceramides in this respect. It is also noteworthy that incorporation of DAG or ceramides into PE bilayers causes the appearance of 'shoulders' or 'bumps' in the transition endotherm. These are probably reflecting inhomogeneities of DAG (or ceramide) distribution, leading to regions of phase coexistence in the phase diagrams.

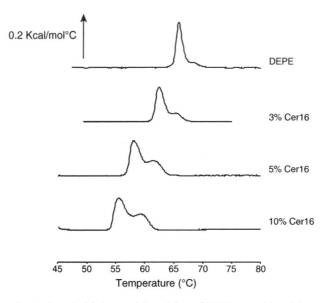

Figure 4.6 Lamellar-to-inverted hexagonal transition of DEPE–ceramide mixtures. Ceramide was *N*-hexadecanoyl sphingosine (Cer16). Data from J. Sot (unpublished)

Box 3   Human platelet activation by cold

Platelets are essential in haemostasis (stopping bleeding). When haemorrhage (bleeding) occurs, the circulating platelets are 'activated' as a result of a complex reaction cascade, and they contribute to form the clot that will stop bleeding. It has been known for a long time that platelets are spontaneously activated when stored at 4 °C, with the implication that, in blood banks, platelets cannot be refrigerated, in order to avoid (irreversible) activation. In fact, cold activation of platelets occurs because their membranes have a single, surprisingly cooperative, transition, at about 15 °C. The thermotropic transition from the fluid to the gel state triggers the mechanisms of platelet activation. Both the plasma membrane and the so-called 'dense tubular system' (inner membranous system) of human platelets have been isolated and studied by DSC. Thermotropic phase transitions in the two membrane fractions were found to occur at the same temperature ($\sim$15 °C) as the intact platelets.

4.6   Complex systems: lipid–protein mixtures and cell membranes

4.6.1   Lipid–protein systems

Classical studies carried out independently by Dennis Chapman and Demetrios Papahadjopoulos in the 1970s indicated that membrane proteins modified in specific ways the gel–fluid transitions of phospholipids (Figure 4.7). Intrinsic (or integral) membrane proteins removed, even at low concentrations, the pre-transition of saturated phosphatidylcholines. The main transition was progressively broadened, and ΔH was correspondingly decreased, with little or no change in T_m. In contrast peripheral (or extrinsic) proteins were found not to decrease, or even to increase ΔH, while simultaneously inducing shifts in T_m, often to higher temperatures. This was interpreted as intrinsic proteins producing a disordered state of the lipid chains, intermediate between that of the gel and liquid-crystalline states, while the extrinsic proteins would be interacting mainly with the phospholipid polar headgroups.

These results have been extended and confirmed by numerous further studies. When intrinsic proteins interact with lipid mixtures that are partially immiscible, e.g. 1-palmitoyl-2-oleoyl PC (POPC) and POPE, the calorimetric data suggest that the packing defects at domain boundaries could facilitate the direct interaction of the transmembrane regions of integral proteins with the lipid hydrophobic chains. Thus lipid domain formation would help in the membrane insertion of proteins.

Ronald N. McElhaney and co-workers have refined the DSC studies of the effects of intrinsic proteins on the gel–fluid transition of phospholipids, using synthetic

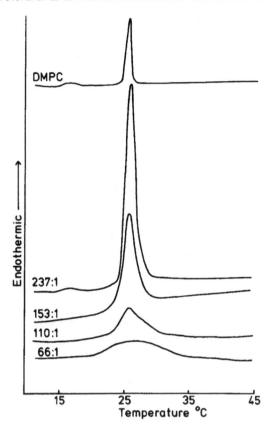

Figure 4.7 Gel–fluid transition of pure DMPC and of DMPC bilayers containing reconstituted sarcoplasmic reticulum C_a^{2+}-ATPase. Lipid–protein mole ratios are indicated for each thermogram. The amount of lipid is the same in all samples, but for the pure DMPC sample instrument sensitivity was decreased by 4-fold. [Reproduced from Gómez-Fernández et al. (1979) *FEBS Lett.* **98**: 224–228, with permission]

peptides that are known to insert with a transmembrane orientation and α-helical conformation, thus mimicking the behaviour of many intramembranous domains of intrinsic proteins. These authors observe that, at low peptide concentrations in saturated PC bilayers, the DSC thermograms exhibit two components, one of which is narrow, highly cooperative, and with properties similar to those of the pure lipid. The T_m of this component, and its fractional contribution to the total enthalpy change ΔH, decreases with increasing peptide concentrations, in a manner that is independent of phospholipid acyl chain length. The second component is very broad, and predominates at high peptide concentrations; the width of this broad transition increases with decreasing acyl chain length. The narrow and broad components have been assigned to the order–disorder transitions of peptide-poor and peptide-enriched bilayer regions, respectively. ΔH of the broad component does not decrease to zero even at high peptide concentrations, suggesting that these transmembrane peptides never abolish completely the gel–fluid transition.

Many other peptides whose mode of interaction with the lipid bilayer differs from that of the transmembrane α-helices, e.g. cyclic antibiotic peptides, have also been studied by DSC from the point of view of their interaction with saturated PC. Not unexpectedly, several different patterns of thermogram alteration have been obtained, witnessing the varying ways of interaction of those peptides with the phospholipid bilayer.

Studies on the effect of proteins on the L_α–H_{II} transition are more recent and less abundant than those devoted to the gel–fluid transition that we have just described. The L_α–H_{II} transition is exquisitely sensitive to the presence of proteins. Small heat-shock proteins (HSP17 and α-crystallin) have been shown to stabilize the liquid-crystalline phase of DEPE, increasing the T_h temperature. A similar effect, in this case with dipalmitoleoylphosphatidylethanolamine bilayers (DPoPE), was found for the synthetic, amphipathic α-helical peptide MSI-78, which has been designed as a cell lytic antimicrobial peptide. These data indicate that both the heat-shock proteins and MSI-78 induce positive curvature strain (the positive sign is arbitrarily attributed to the curvature of the outer monolayer of a cell plasma membrane, i.e. with the fatty acyl chains pointing towards the concave side) in lipid bilayers, i.e. they oppose the change in curvature that accompanies the L_α–H_{II} transition. Conversely pardaxin, another anti-microbial peptide but this one isolated from a fish, lowers the L_α–H_{II} transition in POPE bilayers at peptide–lipid mole ratios as low as 1:50000. Thus pardaxin appears to exert its lytic effects by favouring the formation of nonlamellar structures in membranes.

4.6.2 Cell membranes and cell walls

As was mentioned in previous sections, mixing lipids, and particularly sterols, often has the effect of broadening the phase transitions. Intrinsic proteins have the same effect. Thus it is not surprising that cell membranes, containing highly complex mixtures of lipids and proteins, do not usually exhibit measurable thermotropic phase transitions. With a few exceptions, transitions in cell membranes can only be seen with high-sensitivity DSC instruments.

However the first, and perhaps most significant, studies of DSC on cell membranes were carried out with conventional DSC instrumentation, more than 30 years ago, by R.N. McElhaney and co-workers. The first membranes under study were obtained from the simple bacterium *Acholeplasma laidlawii*, that lacks a cell wall. This microorganism can incorporate substantial amounts of exogenous fatty acids (pro-vided with the growth medium) into its membrane lipids. When these membranes were analysed by DSC, two endothermic transitions were observed, of which the broad, lower-temperature endotherm was attributed to the membrane lipids. This transition was found to vary in temperature in the expected manner with changes in chain length and unsaturation of the fatty acids added to the growth medium. Although *A. laidlawii* cannot synthesize sterols, it does incorporate cholesterol into its membranes from the growth medium, in which case the membrane lipid transition is abolished. Using *A. laidlawii* mutants deficient in enzymes of fatty acid biosynth-esis, it has become possible to prepare bacterial membranes containing essentially a single fatty acid species, in which case the width of the thermotropic phase transition could be reduced from a value of 25–30 °C in natural membranes, to ~7 °C, showing that fatty acid heterogeneity is the main cause of broad phase transitions in native membranes. More recently *A. laidlawii* membranes biosynthetically enriched in tridecanoic acid (C13:0) were found to possess an unexpectedly high T_m, yet one that at the growth temperature (37 °C) will permit a predominantly fluid state in the membranes. The studies on *A. laidlawii* membranes are crucial in demonstrating the validity of the model membrane DSC data when applied to the interpretation of *in vivo* physiological events.

Observation of lipid endotherms in eukaryotic cells is usually hampered by the presence of sterols. However, rat liver microsomal membranes, which are relatively low in cholesterol, and in particular mitochondrial inner membranes, which are virtually sterol-free, have been studied by DSC, and in both cases broad transitions have been found, centered at or near 0°C. This shows that, at physiological temperatures, these membranes are in a completely fluid state.

Cell walls contain a lipid fraction in their composition, and, in the particular case of cell walls isolated from *Mycobacteria* (the bacteria causing tuberculosis), a lipid thermotropic transition may be observed between 30 and 65 °C, i.e. at temperatures mostly above the growth temperature. The endotherm is attributed to an unusual group of lipids, the mycolic acids, that are part of these cell walls. Apparently a significant portion of the lipids exist in a low-fluidity state in growing cells. These lipids would be located in the inner leaflet of the cell wall. Further studies have shown again that the chain length of the mycolic acid plays a crucial role in determining the fluidity (thus the permeability) of mycobacterial cell walls. Moreover, a mutant *Mycobacterium* has recently been isolated that is defective in the synthesis of mycolic acids, and whose cell walls do not exhibit the large thermal transitions typical of the native mycobacterial cells.

4.7 Conclusion

DSC is widely used in biophysical studies of lipids and proteins. It measures very small amounts of heat (in the order of microcalories) that are taken up by or released from biomolecules as a result of chemical or physical changes. In this chapter we have concentrated our attention on the use of DSC to study physical changes (known as phase transitions) in membrane lipids.

A phase is defined as any homogeneous part of a system which is physically distinct, being separated from other parts of the system by a definite boundary, in thermodynamic equilibrium, e.g. water molecules at 1 atm, between 0 and 100 °C, are in the liquid phase; at lower temperatures water is in the solid phase (ice). Membrane lipids at 1 atm can exist in a variety of phases (e.g. lamellar gel, lamellar fluid, inverted hexagonal). Lipid phase transitions (e.g. from lamellar gel to lamellar fluid) may be induced by adding or removing heat to or from the system (adding heat is required in the specific example of the gel-to-fluid transition). These transitions are said to be thermotropic, i.e. heat-driven. DSC is ideal for measuring the heat exchanges that accompany lipid phase transitions.

Observations of thermotropic phase transitions even in simple lipid mixtures are complicated by factors such as nonideal miscibility, coexistence of different phases under certain conditions, and specific interactions between lipids, leading to complex formation. A well-studied, yet incompletely understood, binary lipid mixture that exemplifies these difficulties is the one formed by phosphatidylcholine and cholesterol.

Membrane proteins have a profound influence on the thermotropic properties of lipid bilayers. Moreover, the effects of intrinsic (or integral) proteins are very different from those of extrinsic (or peripheral) proteins, reflecting their different modes of interaction, respectively, from the hydrophobic and hydrophyllic moieties of lipids. Some proteins can also influence, even at very low protein:lipid ratios, the lamellar to inverted hexagonal transitions of certain lipids.

Structural variations of isolated cell membranes and even, in favourable cases, whole cells can be studied by DSC. In spite of the inherent difficulties, the available data confirm that, in essence, the results obtained from simple lipid–water systems and model membranes can be applied to understand the structure–function relationships of cell membranes.

Further reading

The first block are of general interest. The second block are selected examples of specific DSC applications.

G. Cevc (1993). *Phospholipids Handbook*. Marcel Dekker: New York.

Differential scanning calorimetry (an introductory tutorial); www.psrc.usn.edu/macrog/dsc.htm.

A. Cooper, M.A. Nutley and A. Wadood (2001). Differential scanning microcalorimetry. In *Protein–Ligand Interactions: Hydrodynamics and Calorimetry. A Practical Approach*, Edited by S.E. Harding and B.Z. Chowdhry, pp. 287–318. Oxford University Press: Oxford.

G.W.H Höhne, W.F. Hemminger and H.J. Flammersheim (2003). Differential Scanning Calorimetry: An Introduction for Practitioners, 2nd ed. Springer: Berlin.

Lipidat (a database of physical properties of lipids); www.lipidat.chemistry.ohio-state.edu.

A.G. Lee (1977). Lipid phase transitions and phase diagrams. *Biochim. Biophys. Acta* **472**: 285–344.

McElhaney, R.N. (1982). The use of differential scanning calorimetry and differential thermal analysis in studies of model and biological membranes. *Chem. Phys. Lipids* **30**: 229–259.

D. Chapman, J. Urbina and K. Keough (1974). Biomembrane phase transitions. Studies of lipid–water systems using differential scanning calorimetry. *J. Biol. Chem.* **249**: 2512–2521.

X.L. Cheng, Q.M. Tran, P.J. Foht, R.N. Lewis (2002). The biosynthetic incorporation of short-chain linear saturated fatty acids by *Acholeplasma laidlawii* B may suppress growth by perturbing membrane lipid polar headgroup distribution. *Biochemistry* **41**: 8665–8671.

A. Ortiz, J. Villalain and J.C. Gómez-Fernández (1988). Interaction of diacylglycerols with phosphatidylcholine vesicles as studied by differential scanning calorimetry and fluorescence probe polarization. *Biochemistry* **27**: 9030–9036.

J. Sot, F. J. Aranda, M. I. Collado, F. M. Goñi and A. Alonso (2005). Different effects of long and short-chain ceramides on the gel-fluid and lamellar-hexagonal transitions of phospholits. A calorimetric, NMR and x-ray diffraction study, *Biophys. J.* **88**, 3368–3380.

5

Membrane potentials and membrane probes

Paul O'Shea

Cell Biophysics Group, The School of Biology, University of Nottingham, NG7 2RD, Nottingham, UK

5.1 Introduction: biological membranes; structure and electrical properties

The fluid-mosaic model of cellular membranes illustrated in Figure 5.1 helped enormously in devising further experimental investigations of the properties of membranes. Some of these then led to elaborations of the general structure as outlined in Figure 5.1. A commonly held view of (perhaps) the sole function of membranes, however, is that they act to compartmentalise soluble components (from ions to enzymes) of cells by acting simply as a selective permeability barrier. Although, this viewpoint is by no means incorrect, to consider it the singular role of membranes is rather too simplistic; membranes possess properties that underlie subtle and highly sophisticated additional modes of behaviour. These include a repertoire of different electrical potentials and indicate that in many ways the membrane represents a unique molecular environment in its own right. These electrical properties will be outlined with attempts to suggest their key roles but with the main emphasis on their measurement.

Three membrane potentials appear to be associated with cellular membranes but can also be present in artificial membranes. The cellular membrane potential that has received the most experimental attention is associated with a gradient of electrical charge across the phospholipid bilayer known as the *transmembrane potential difference* (with the symbol $\Delta\psi$). These charge gradients are engineered and

Chemical Biology   Edited by Banafshé Larijani, Colin. A. Rosser and Rudiger Woscholski
© 2006 John Wiley & Sons, Ltd

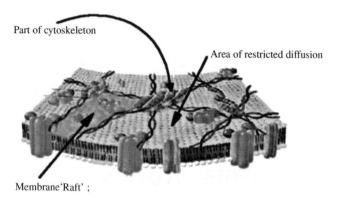

Part of cytoskeleton

Area of restricted diffusion

Membrane 'Raft' ;

Figure 5.1   Contemporary view of biological membrane

maintained (with an energetic cost) by the translocation of charged solutes (eg. Na^+, K^+ or H^+) or electron transport across the membrane. Of all the membrane potentials, the biological and physical roles of the $\Delta\psi$ are the best documented. The second 'membrane potential' of major significance is known as the *membrane surface potential* and its existence has been established for some time with a well developed and robust underlying physical theory. Finally, the third manifestation of a 'membrane potential' has been described as the *membrane dipole potential* as it results from the contribution that molecular polarisations or electrical dipoles make to the properties of biological membranes. These three 'membrane potentials' are outlined in greater depth in the recently published volume *Bioelectrochemistry of Membranes* (see Further reading) but some aspects of their properties will be outlined to indicate their evolving importance.

5.2   Phospholipid membranes as molecular environments

The fluid-mosaic structure of membranes offers a relatively fluid structure permitting translational molecular diffusion in 2D and semi-permeability in the third dimension. Much basic knowledge relating to this was accumulated throughout the 1970s and 1980s as well as an enormous expansion of information concerning how signalling reactions take place at and within membranes. There had been debates about experimental inconsistencies for some time and much progress was made in identifying the biophysical rationale for this. Throughout the 1990s however, there was a growing appreciation that molecular (particularly macro-molecular) diffusion within membranes was not as simple as first envisaged and this pointed to a more sophisticated membrane organisation. Much of this resides in recently defined membrane microdomain structures sometimes (but not always) known as *Rafts*.

These assemblies are thought to be stabilised by the physical interactions that underlie the phenomenon of membrane phase-separation elicited by sterols such as cholesterol and sphingolipids in the membrane bilayer but proteins appear also to be involved. Rafts seem to exist as viscous 'patches' rich in sterols and sphingomyelin 'afloat' in the much more fluid 'Singer & Nicholson' membrane. Their have been few studies on the structure of rafts but some evidence exists that they exhibit structural heterogeneity. Rafts are believed to 'recruit' glycosylphosphatidyl-inositols (GPI) and GPI-linked proteins. Rafts appear to be of major biological importance as their attendant (GPI-linked) proteins play many significant roles.

5.3 The physical origins of the transmembrane (V_m or $\Delta\psi$) surface (ϕ_S) and dipolar (ϕ_D) membrane potentials

A number of quite different techniques have evolved to make measurements of the various electric potentials associated with membranes but particular measurements must rely on the physical differences between the potentials. In order to avoid interpretative difficulties due to interference between each potential, clearly it is necessary to have a very clear understanding of the physical nature of both the potential as well as the measurement technique. In addition, however, membranes also possess other properties which often complicate assignments of specific values for the membrane potentials.

5.3.1 The transmembrane potential difference (V_m or $\Delta\psi$)

Membranes represent formidable barriers to the movement of ions (and electrons). Thus, transport of ions as net charges across the insulator offered by the membrane will establish an electrical potential difference (a voltage $= V$) across the membrane (V_m or $\Delta\psi$). This phenomenon is well known and can be described by Equation. (5.1):

$$V = q/C \quad = \Delta\psi \quad (= \Delta\phi_m) = V_m \qquad (5.1)$$

where, V is the electrical potential difference across the membrane ie. V_m or $\Delta\psi$, $C =$ the capacitance of the membrane (dielectric) and q is the number of charges transported across the membrane. The term, $\Delta\phi_m$ is included as this represents a more consistent nomenclature, and would be better recognised by other disciplines (eg. Electrochemists), however as $\Delta\psi$, is used widely in the biological literature, it will be utilised in the present chapter.

The transport of electrical charge across a membrane may take the form of cation, anion (both inorganic and organic) or electron transport. It can be passive due to membrane leaks (either inherent or facilitated by protein channels, ionophores, detergents etc). In these latter cases the transport is driven by the difference in electrochemical potential of the particular ion between the two phases separated by the membrane. In biological membranes, however, the charge transport is typically

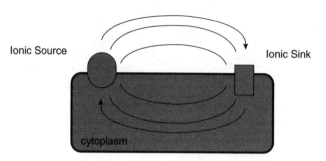

Figure 5.2   Ionic circuits in living systems

coupled either to the movements of other ions due to co- or anti-porting enzymes, or to chemical reactions as in the case of oxidative phosphorylation, absorption of light energy (ie. photosynthesis) or hydrolysis of ATP. This is illustrated in Figure 5.2 below in which the intracellular milieu labelled as 'cytoplasm' could be eukaryotic or prokaryotic or in fact, any of the intracellular compartments associated with organelle/plastid/vesicle systems.

The value of $\Delta\psi$ established in such complex systems is governed by the relation

$$\Sigma_i Z_i J_i = 0 \qquad (5.2)$$

Equation (5.2) implies that when summation of all the fluxes (J_i) of charged species (Z_i) across the membrane leading to charging of the membrane capacity has concluded a quasi-equilibrium with a (approximately) constant V_m or $\Delta\psi$ is attained. Its value depends on the difference in chemical potential $(\Delta\mu_i)$ of all transported species and on the affinities of coupled chemical reactions. This is embodied in the so-called Goldman diffusion equation but these relations are essentially transcendental with their explicit solution tractable only under defined conditions. If only one species permeates through the membrane, for example, a true equilibrium state is reached and the resulting transmembrane electrical potential difference is described by an expression known as the Nernst equation

$$V_{\mathrm{m}} = \Delta\psi = (\phi_o/Z_i)\ln(c_{i,1}/c_{i,2}) \qquad (5.3)$$

with the abbreviation

$$\phi_o = kT/e = RT/F \qquad (5.4)$$

R and F are the gas constant and the Faraday constant, respectively, and $c_{i,1}$ and $c_{i,2}$ refer to the concentrations of the species in each of the two aqueous phases separated by the membrane with an assumption of equal values of μ_i^0 in both phases. Unfortunately the nomenclature of Eqn. (5.3) can be confusing as another well known relation is also referred to as the Nernst equation. This relates the oxidation-reduction (redox) potential to the concentrations of the components of a redox couple. The redox potential is not a transmembrane electrical potential difference but is measured between an inert metal and a reference electrode, both in the same phase and represents the chemical potential of the electron.

5.3.2   The membrane surface potential (ϕ_S)

This becomes manifest when a net excess electric charge resides at the membrane interface in contact with the surrounding aqueous medium. Thus an electrical potential difference exists between the surface of the membrane and the surrounding aqueous bulk phase milieu leading to the so-called the electrical (Helmholtz) 'diffuse layer'. This potential 'reaches out' from the membrane surface into the aqueous bulk phase as illustrated in Figure 5.3. In fact several such layers adjacent to the surfaces

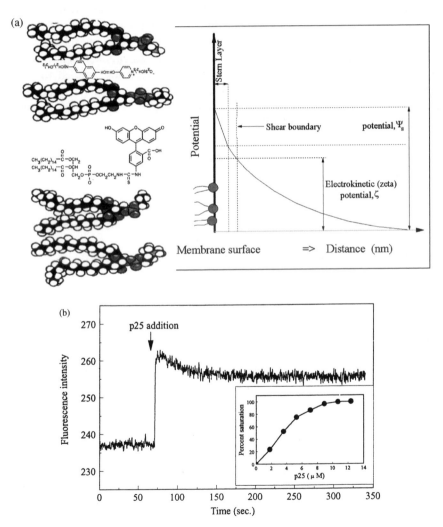

Figure 5.3 (a) The surface potential versus distance; (b) The use of FPE to measure macromolecule-membrane interactions

of membranes are thought to exist and are essentially analogous to those defined by electrochemists in the context of the electrode-solution interface. Care must be taken when simply reworking concepts that have their origins in Electrochemistry. There are important differences between membranes and electrodes that may complicate formal descriptions of the properties of the former, particularly in relation to how they manifest in a biological context. Electrodes are essentially homogeneous hard, solid metal surfaces, whereas membranes are soft, fluid, highly heterogenous interfaces that interact strongly with and through the 'hydrophobic effect' actually 'rely' on their aqueous environment to exist. This is perhaps exemplified best by the presence of the dipole potential, a property for the most part absent in electrodes but enormously influential in the membranes of living cells (see below).

Surface potentials at the electrode-solution interface have been described by a number of formalisms. The most successful of these originates from Gouy and Chapman whom suggested that Poisson-Boltzmann approaches best describes the state of affairs at the electrode surface in contact with an aqueous solution (further elaborations are outlined by Bockris & Reddy). Within Electrochemistry this proved very successful and analogous formalisms were subsequently applied to physical descriptions of biological surfaces. The resultant Poisson-Boltzmann equation with defined boundary conditions can be solved analytically to yield an expression for the membrane surface potential as follows:

$$\phi_s = (2\phi_o/Z)\mathrm{Arsinh}[Z\phi_\sigma/(2\phi_o)] \tag{5.5}$$

with ϕ_o as in Eqn. 5.4 and the abbreviation $\phi_\sigma = \sigma\lambda/(\varepsilon_r\varepsilon_o)$ whereby σ represents the surface charge density, and λ is known as the Debye length. At 25°C $[\varepsilon_r\,\varepsilon_o\,kT/(2\,e^2N_A)]^{1/2} = 0.304\,\mathrm{nm}\,M^{1/2}$ for an aqueous phase with a dielectric constant $\varepsilon_r \approx 80$.

The potential profile in the diffuse layer follows as:

$$\phi(x) = \phi_o \ln\{[1 + \tanh(\phi_s/2\phi_o)\exp(-x/\lambda)]/[1 - \tanh(\phi_s/2\phi_o)\exp(-x/\lambda)]\} \text{ for } \quad x \geq 0 \tag{5.6}$$

The distance-dependence of the surface potentials depends on a number of parameters such as the dielectric constant, the ionic strength, the net excess surface charge density, temperature etc. An indication of the order of magnitude and distances involved of this is shown in Figure 5.3.

These expressions however are essentially, mean-field approaches as little account was taken of the molecular diversity that exists in living cell membranes. This is in part due to the evident complexity of such membranes but some modelling simulations have now begun to incorporate such diversity (eg. lipid-protein interactions are emphasised within the context of computational modelling). These are important tasks as there is clearly much local variation in the magnitude of the membrane surface potentials. We have engaged this problem for some years and are able to visualise the spatial variation of the potential about the membrane and cell surface with high molecular and spatial resolution (see below).

5.3.3 The membrane dipole potential (ϕ_d)

This originates from the dipolar configurations of the lipidic components of the membrane bilayer, polar moieties such as $C^{\delta+}O^{\delta-}$ and $O^{\delta-}P^{\delta+}$ etc. and ordered water molecules at the molecular surface of the membrane also make a contribution. The organisation of the membrane components that contribute to this potential have been verified from neutron diffraction studies and NMR spectroscopy. These dipolar groups appear to be oriented in a way such that the potential located towards the hydrophobic interior of the membrane is positive with respect to the pole located towards the external aqueous phases, and ϕ_d has a magnitude of several hundred millivolts (typically about 300 mV). Formalising this for analytical use, however, has not been straightforward to implement and some of the discussion below outlines some of the problems.

Symbolically formalising the measured molecular arrangements in order to undertake modelling of the membrane dipole potential is also not as straightforward as it may seem. It is possible to begin by identifying a vector drawn from the point of the negative charge $-Q$ to the positive charge $+Q$ of any dipole which is the familiar electric dipole moment $\mathbf{p}$. The magnitude is $Q\,a$, with a defining the distance between the centres of the charge density. The potential at a given point with a position vector $\mathbf{r}$ with respect to the centre of the dipole can be expressed as:

$$\phi = \mathbf{p}\,\mathbf{r}/(4\pi\varepsilon_r\varepsilon\,\mathbf{r}^3) = p\,\cos\theta/(4\pi\varepsilon_r\varepsilon\,\mathbf{r}^2) \qquad \text{for} \quad \mathbf{r} \gg a \qquad (5.7)$$

where θ is the angle between $\mathbf{r}$ and $\mathbf{p}$. Higher order terms become influential if $\mathbf{r} \gg a$ is not fulfilled.

Equation (5.7) would be an enormous simplification if it is all that is required to describe the membrane dipole potential but a number of non-trivial complications need to be included in any such analysis. The solvent environment for example, is often dealt with as a mean field or in a continuum manner. But the relative permittivity (or dielectric constant) (ε_r) cannot be considered to possess the same value throughout the multiphase system represented by a membrane in an aqueous medium. The permittivity profile has been measured to vary from about 78.5 in the bulk aqueous phase to 20–30 in the diffuse layer at the interface then to around 2 in the membrane interior. Furthermore, in light of the discussion above concerning the complexity of the membrane-solution interface region, it should be born in mind that there are certainly more than 3 such distinct phases!

One way in which to deal with this for protein modelling has been to use an *ad hoc* 'continuum' or mean field approach in the form of a 'distance-dependent dielectric constant' in which the dielectric constant is taken to be proportional to separation. Thus dielectric screening is weak at small separations and larger as the separation increases. Although this approach requires a relatively small computational overhead it leaves much to be desired and explicit solvent models would be far more preferable. Equation (5.7) would require significant elaboration, however, to include these latter complexities.

A related problem that also adds another layer of complexity is the fact that even if it is possible to write an elaborated equation according to the foregoing complications above that are necessary to deal with the multi-parameter environment, within the body of a membrane there are likely to be many different lipid types each with their own set properties and explicit dipole configurations. These latter points therefore, underline the initial comments in the introductory sections to this chapter that a membrane must not be considered simply as an homogeneous slab of low dielectric insulating material.

Finally, it is evident from Equation 5.7 that the dipole 'decays' as $1/r^2$ rather than the $1/r$ dependence of an isolated point charge. Accordingly the magnitude of the dipole potential decreases steeply during passage from the water/membrane interface into the body of the membrane (ie. as r^{-3}). The resulting forces on another polarised body would decline just as steeply (ie. the well-known $1/r^6$ dipole-dipole interaction), however, as it moved more deeply into the body of the membrane, until it encounters the dipoles located towards the other bilayer leaflet.

The foregoing points would not really be important issues however, if the dipole potential simply was a phenomenon that exists without any clear role in cell biology. This is not the case however, for as commented above, we have pioneered the concept that local dipolar fields appear to have profound effects on membrane function not least in modulating how molecules interact both with and within membranes. Furthermore, and perhaps even more importantly because it appears to modulate membrane protein structure and thus function.

5.4   Measurement of membrane potentials

5.4.1   Electrodes

Perhaps the most obvious tool for measuring electrical potentials involves the use of electrodes; ie. comparison of the potential at a measuring electrode with that of a reference electrode yields some estimation of the potential in question. Penetration of a membrane (such as that of an organelle or a plasma-membrane patch) when compared to a reference electrode potential, for example, facilitates an estimation of the trans-membrane voltage (ie. V_m or $\Delta\psi$). Despite the practicality and reliability as well as the great sensitivity of techniques such as patch-clamping, related techniques offer less assured interpretations. There are many reasons for example, why fully penetrating a plasma membrane with an electrode is unsatisfactory because even micro-electrodes compared to the size of a single cell are essentially large objects, thus penetration of a cell by the electrode may actually perturb the value of the measurement (eg. by affecting cellular metabolism which underlies the transmembrane potential). Similarly, measurement of V_m, with microelectrodes or the whole-cell current-clamp technique have the drawback that the recording pipette is in contact with the cytoplasm, and dialysis with the pipette solution may perturb the ionic composition of the interior of the cell.

Measurements of the surface potential may be similarly undertaken in principle, by bringing an electrode very close to the surface of a membrane (ie. see Figure 5.3) and comparing the measured potential to that of a remote reference electrode. The remaining 'electrode' based method of making measurements of membrane surface potentials involves passing a current between electrodes and determining the rate of movement of membranous particles. The current generates an electric field which acts on membrane-located charges and leads to bulk electrophoresis of the particles. In order for this technique to be applied, the membrane system must exist as a relatively homogenous particulate suspension but may also be used routinely with large phospholipids vesicles (ie. usually multi-lamellar liposomes) or with human cells (eg. see Table 1). By application of the Helmholtz-Smoluchowski equation (see references below), the quantity known as the *zeta* potential (ζ) (this is a less precise expression of the surface potential and referred to as the electrokinetic potential) is obtained. This bears some relation to the surface electrostatic potential (see Figure 5.2), although it is not formally well defined and possess' some interpretative difficulties. Nevertheless, particle microelectrophoresis has been implemented in a number of membrane systems including my own laboratory, to make measurements of the binding of macromolecules to human cells (Wall *et al.* 1995). The results appear to be consistent and reliable but the nature of the particle electrophoresis technique does not permit its application with adherent cells. This is a major shortcoming, as is the fact that rapid time resolution of these '*surface potential*' changes is also not possible. The latter is a limitation because these techniques require that the surface is in electrical equilibrium over the time scales of the measurement; often many measurements are necessary each taking seconds. This leads to interpretational problems for if further changes at the cell surface take place following ligand binding (eg. as with receptor-mediated endocytosis) then the particle electrophoresis approach may be misleading.

Table 5.1 indicates measurements of the zeta potentials of cells in various media and in the absence and presence of representative types of protein which are anticipated to bind to the cell surface. The data in Table 5.1 also show the anticipated effects of electrolytes on the zeta potential of human cells. There are reports of variations of the value of the zeta potential of blood cells and we have found in our studies that erythrocytes possess differing amounts of sialic acid which results in

Table 5.1 Zeta potential measurements of human cells: effects of medium and serum albumin

Cell type	Suspension Medium	Zeta Potential (to nearest mV)
Lymphocytes	Iso-osmotic-Low ionic strength	−27
	Physiological ionic strength	−18
Erythrocytes	Iso-osmotic-Low ionic strength	−40
	Physiological ionic strength	−19
	Physiological ionic strength $+ 2.2\,\mathrm{mg\,ml^{-1}}$ albumin	−65

Zeta potentials were calculated on the basis of the observed electrophoretic mobilities of the cells in the respective media. For each determination, not less than 20 different measurements of each cell type were taken.

different surface potentials. The origin of these variations, in a biological context however, are potentially interesting as diet, metabolism and a number of pathological conditions seem to underlie the variation of the surface potential. The functionale rationale of this therefore, may reside in observation that the ability for erythrocytes to self-aggregate appears to be modulated depending upon the nature of the sugar carbon source. Thus under circumstances of uncontrolled diabetes, a less electro-negative surface potential results leading to a smaller coulombic repulsion between cells predisposing a greater possibility for aggregation and hence thrombosis.

Finally, measurements of the dipole potential may also be accessible utilising electrode technology. Brockman (1994) has reviewed several such applications but it must be conceded that they too, leave much to be desired.

The most recent technological development involving the application of what could be described as an 'electrode' towards membrane studies involves the use of the scanning tunnelling microscope or more recently, the scanning probe and Atomic Force microscopes (AFM). With this instrument, it proved possible to map the surface morphology and also it is feasible to measure the electrostatic nature of the surface at varying distances from the surface. Thus, quantitative estimations of the surface potential may be attempted; it is not clear, however, whether the dipole potential may also have an effect on this value. Even ignoring this latter possibility that the dipole potential may well interfere with such estimations, the $\Delta\psi$ across a membranous structure is not accessible using AFM technology. Nevertheless, this is one of the few emerging techniques that offers spatial information about any of the membrane potentials (although see below). The main problem related to spatial measurements however, has been that the single AFM probe would have to interrogate the whole surface area of the membrane region of interest. Thus, the rastering nature of technique would add a time requirement for measurement that would be unacceptable. Lately however, much more rapid rastering techniques have been developed that facilitate video-rate imaging of a cell surface feasible (M. Miles, Bristol, UK, personal communication). Other problems less likely to be so 'easily' circumvented include the fact that no intra-cellular measurement is possible with external scanning probe approaches and interference of the extra-cellular matrix or glycocalyx of living cells means that extra-cellular scanning would also be problematical!

Other than developments of AFM as directed towards cell surfaces, the remaining less-sophisticated applications of electrodes do not offer satisfactory approaches to determining membrane potentials not least because of the continuing drive towards collecting spatial information. Presently, therefore, the most widely used techniques make use of spectroscopic approaches; in both invasive and non-invasive applications. Of the latter, NMR spectroscopy has been applied to interrogate membranes in many different ways and has yielded an enormous amount of valuable information about the structures of both lipid and protein membrane components. More particularly, NMR has been used recently to develop a clearer understanding of the nature of the membrane dipole-potential and actually two decades ago as a tool to monitor the membrane surface charge. And whilst certainly not criticising the unparalleled achievements of NMR-approaches to macromolecular aspects of cell biology, it must be conceded that NMR suffers from low sensitivity and poor spatial resolution. Thus it is not the tool of choice for single cell biology. Whilst, NMR techniques directed towards membranes have been either totally non-invasive or effectively non-invasive by isotopic-substitution of appropriate atomic nuclei, a number of other techniques rely on the inclusion of optical probe molecules to report the membrane properties. Numerous spectroscopic membrane probes have been developed to study a number of membrane properties. These technologies are often fairly simple to use and to be very versatile. Studies have been performed using many different optical intrinsic or extrinsic probes which exhibit potential-dependent properties (Molecular Probes Catalogue Inc.). The majority of these tend to take advantage of fluorescence rather than optical absorbance mainly because fluorescence is so much more sensitive, versatile and yields much better signal to noise detection.

5.4.2 Spectroscopic measurements of the transmembrane potential difference

This area of study is very well established and there are many techniques and spectroscopic probes available (eg. Bunting *et al*. 1989). Originally the strategy was to assess the redistribution of membrane-permeant indicators that migrate according to the trans-membrane potential difference as embodied in the Nernst equation. Thus transmembrane potential-dependent accumulation of a fluorophore such as Rhodamine 123 into mitochondria reports the membrane potential as variations of fluorescence (eg. Bunting *et al*. 1989). This takes place according to the Nernst equation (see above) and through this agency the signals may be standardised for different experimental systems. Calibration of the signal response can be achieved, therefore, with the aid of defined transmembrane potentials generated by K^{+}-diffusion potentials and a K^{+}-specific electrogenic ionophore such as valinomycin. More conveniently other fluorophores may possess electrochromic properties and report the membrane voltage directly. It is clear, however, that one added advantage of using spectroscopic indicators that respond to the transmembrane potential difference, was that they also offer the possibility of obtaining kinetic information of the changes of potential due to physiological circumstances.

Other electrochromic membrane probes of note include, merocyanine 540 used to study the transmembrane potential in liver mitochondria but some problems are also evident. Although membrane potential measurements in mitochondria have been identified in the foregoing section, many of these strategies are generic and hold for plasma membranes of (eg.) neurones or other excitable tissues, as does the following in which more cellular studies are emphasised.

Overall, measurements of transmembrane potential differences in living cells is fairly well established with the methods fairly reliable and robust. These mostly take on two strategies either utilising electrodes or optical tools such as fluorescence to record the potential differences (V_{m} or $\Delta\psi$) based on electrochromism or probe accumulation according to the Nernst equation (5.3).

5.4.3   Spectroscopic measurements of the membrane surface potential

Amongst the techniques developed for obtaining electrostatic details of the membrane surface using spectroscopic probes applications with two categories of probe seem to have evolved. Probe molecules either redistribute themselves or change their spectral properties according to the magnitude of an electrical potential at the membrane surface. One of the main criticisms for these approaches, however, resides in problems of the uncertainty as to the exact location of the probes during the course of measurements. We have addressed many of these problems and introduced the use of a family of fluorophores attached to phospholipids. These have the advantage of being virtually non-invasive as they are used at very low concentrations, do not perturb the membrane and can be easily and precisely located.

One such probe molecule, Fluorescein Phosphatidyl Ethanolamine (FPE), has proved to be a versatile indicator of the electrostatic nature of the membrane surface in both artificial and cellular membrane systems. FPE is sensitive to changes in the surface potential ϕ_s at the membrane-solution interface because the fluorescent moiety of the FPE lies precisely at the membrane solution interface as illustrated in Figure 5.3. Any changes in the number of surface charges at the membrane, such as the binding of an inorganic ion or a charged oligopeptide, will cause an alteration in ϕ_s. These probe molecules sense this because according to the Boltzmann equation the concentration of protons at the membrane surface is

$$c_{i,s} = c_{i,b} \exp[-Z_i F\phi_s/(RT)] \tag{5.8}$$

thus when introduced into the logarithmic form of the Henderson-Hasselbalch equation, yields upon rearranging:

$$\log(c_B/c_{HB}) = pH - [pK - F\phi_s/(RT \ln 10)] \tag{5.9}$$

Thus $pH = -\log(c_{H,b})$, while c_B and c_{HB} represent the concentrations of the dissociated and protonated species of the acid-base chromophore pair, respectively. The quantity $pK - F\phi_s/(RT \ln 10)$ can be considered as an apparent pK for proton binding of the chromophoric acid-base couple on the membrane surface. Eqn. 5.8 shows that the protonation state c_B/c_{HB} of the probe is altered if ϕ_s changes at constant pH, which results in a change in the fluorescence yield. Thus changes of fluorescence at constant bulk pH reflect changes of ϕ_s and can be utilised, therefore, to study how molecules interact with membranes.

Any changes in the number of surface charges at the membrane, such as the binding of an inorganic ion or a charged oligopeptide, will cause an alteration in ϕ_s, in accordance with the mode of action of the probes (Wall *et al.* 1995), and thus results in a change in the fluorescence yield. This technique has been utilised to measure the time course of the interactions of charged molecules such as Ca^{2+}, peptides, and proteins with synthetic and biological membranes with great sensitivity. By performing rapid-kinetic studies, it has also proved possible to monitor the early events during the interactions of oligopeptides with artificial membrane systems and with cells. Thus the phenomenon may be utilised to measure the time course of the interactions of virtually any charged molecules also with biological membranes with great sensitivity and in real time. An example of this is shown in Figure 5.3 in which an leader or signal oligopeptide (known as p25) is observed to interact in real-time with well-defined phospholipid membranes.

Estimations of thermodynamic information of the various components of the interactions (ie. binding and insertion with clearly separate kinetic rate constants) are possible which are not easily obtained by other means (Cladera & O'Shea 1998). One of the virtues of this approach is that virtually identical studies may be performed with living cells so direct comparisons of the interactions can be undertaken.

5.4.4 Spectroscopic measurements of the membrane dipole potential

Since the possibility that the dipole potential may have some influence on cellular function was the last to be appreciated, techniques designed to make measurements have not become either as well-established or as varied. Nevertheless, experiences with measurement strategies for the other potentials has meant that measurements of dipole potentials are at least as sophisticated as those used for other membrane potentials. The dipole potential ϕ_d has been measured using a class of potentiometric fluorescent indicators that operate by electrochromic mechanisms. Di-8-ANEPPS [1-(3-sulfonatopropyl)-4-[β[2-(di-n-octylamino)-6-naphthyl] vinyl] pyridinium betaine] in particular has been successfully applied to the measurement of ϕ_d using dual-wavelength ratiometric fluorescence methods. This method forestalls problems arising from small differences in dye concentration between different samples, dye bleaching or the influence of light scattering on the fluorescence measurements.

The excitation spectrum of di-8-ANEPPS is altered when it lines up with the phospholipid/sterol dipoles and leads to electronic redistributions within the probe molecule promoting red or blue shifts in the excitation spectrum depending on the magnitude and direction of the dipole moment of the ambient environment that the probe finds itself in. These phenomena are illustrated in Figure 5.4, in which membranes with various lipids preparations (etc) are prepared and lead to spectral differences.

In our laboratory we have fully characterised this phenomenon to make use of the probe as an indicator of the intermolecular interactions that take place within membranes. Comprehensive studies of the dependence of the relative magnitude of the dipole potential on the membrane lipidic composition have been reported. Of some interest, however, was the observation that di-8-ANEPPS may be used to indicate the interactions of some macromolecules with membranes and we were able to show that the dipole potential also has an effect on the structure of peptides within the membrane. It appears from our findings that the magnitude of ϕ_d is measurably influenced by peptides/proteins that insert (at least partially) into the membrane and *vice versa*. An example of this phenomenon is shown in Figure 5.5.

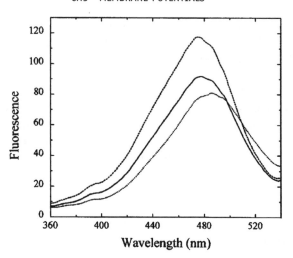

Excitation spectra (λ_{em} = 580 nm) of PC (——), PC-KC (KC 15 mol%) (– – –), and PC-phloretin (phloretin 15 mol%) (⋯⋯) liposomes labelled with 4 μM di-8-ANEPPS. The lipid concentration was 200 μM.

Figure 5.4   Fluorescence excitation spectrum of di-8anepps in membranes made up of PC, PC & 15% Ketocholestanol and PC & 15% Phloretin

The excitation spectra of membranes labelled with di-8-ANEPPS; spectra of PC membranes with those obtained containing 15 mol% of either Ketocholestanol (KC) or phloretin are also compared in Figures 5.4 and 5.5. These sterol compounds promote changes in the membrane dipole potential and significant variations of the intensity and position of the excitation maximum were observed when these sterols are incorporated into the membrane bilayer. Ross *et al.* (1994) have pointed out that such intensity variations may be due in part to a potential-dependent shift in the emission spectrum which gives a change in intensity at the fixed emission wavelength used for the excitation scan. To determine the changes resulting only from the spectral shift, the areas of the excitation spectra must be normalised to the same integrated intensity and then subtracted. This procedure yields the difference spectra shown in Figure 5.5 which were obtained by subtracting the PC-membrane spectrum from the PC-Phloretin membrane spectrum which shows a minimum at 450 nm and a maximum at 520 nm. The presence of KC however, yields a difference spectrum with a maximum at 450 nm and a minimum at 520 nm, which is the opposite effect to that of phloretin. Recording the ratio (R) of fluorescence excited at the two wavelengths with the maximum positive and negative changes provides a method, therefore, for measuring spectral shifts originating from changes of the local electric field and avoiding artifactual variations in intensity. The feasibility of measuring the membrane dipole potential in this manner has been explored in several laboratories. In agreement with these studies an increase of the membrane dipole

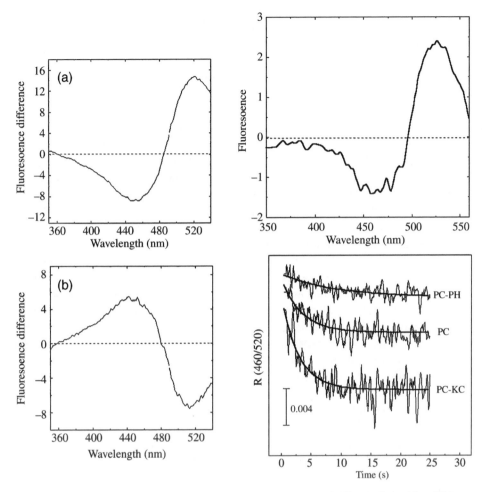

Figure 5.5 Spectral shifts of di-8-anepps; difference spectra and effects of peptide additions

potential occurs due to the presence of KC and a decrease caused by phloretin. Of more interest, however, was the observation shown in Figure 5.5, that di-8-ANEPPS may be used to indicate the interactions of some macromolecules with membranes. The ratiometric spectrum, also shown in Figure 5.5 obtained following the exposure of a membrane labelled with di-8-ANEPPS to oligopeptides yields a spectral shift similar in profile to that obtained with KC. The time course of the interaction of the peptide with the membrane is also shown in Figure 5.5, and correlates well with other studies which indicate that this signal change relates to the insertion and folding of the peptide. The dipole potential, therefore, seems an influential factor in controlling the interaction of the peptide with the membrane. In further studies we demonstrated that the dipole potential has an effect on the secondary structure of the peptide within the membrane.

5.5   Problems with spectroscopic measurements of membrane potentials

All the fluorescence probe technologies outlined above have proved valuable indicators of these most influential parameters associated with membranes. They offer the means of making both measurements of the magnitudes of the membrane potentials and not least also for detecting any changes (slow or rapid) due to molecular interactions within membranes. The most significant problems associated with these technologies, however, are concerned with bleaching of the probes during longer-term (ie. >many mins) interrogations of membranes together with the question of whether the probes themselves may alter any of the membrane parameters under scrutiny. These problems are usually not major issues as photobleaching typically just means there is less active fluorescence probe and so the overall signal declines. If the overall signal intensity is the parameter of interest this can lead to interpretational issues but typically the rate of bleaching is known, can be measured and subtracted from the measurement. On occasions, however photobleaching may also occur leading to undesirable photochemical activity (such as the production of activated oxygen species or chemical reactivity with other membrane components) that may alter the behaviour of the membrane under scrutiny. These are clearly more of a worry but there is usually prior knowledge of these possibilities discovered during the first characterisations of any new probe, so investigators are alerted. As an example of this, lately, we were concerned that doping membranes with di-8-ANEPPS (a molecule that has some structural features reminiscent of eg. KC) may actually augment the population density of 'rafts' within a given membrane that is predisposed to form rafts. Thus such concerns raises the question of the probes reliability in reporting the membrane property we were most interested in. A rigorous analysis however, indicates that this phenomenon has a detectable but minor effect on the raft population and thus the general conclusions outlined in this chapter appear sound. With this in mind however, we have developed additional fluorescence probes that yield similar information about the dipole potential as di-8-ANEPPS and operate essentially in the same manner but without any effect on membrane organisation.

5.6   Spatial imaging of membrane potentials

The last decade or so has seen spectacular increases in applications that reveal 2- & 3-D spatial information of the molecular details of cellular biochemistry. This has already been commented on above concerning the use of AFM but many types (the majority based on fluorescence) of imaging technologies are now routinely applied to biological problems. And given that several of the technologies outlined above that are utilised to determine biological membrane potentials are optically based and particularly involve fluorescence, then spatial imaging is a logical extension of these applications. Many applications of these technologies, therefore, have involved fluorescence microscopy for imaging the transmembrane potential difference but

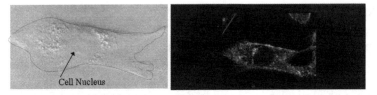

Figure 5.6 Fluorescence microscopy of a single fibroblast labelled with di-8-anepps

this is the only type of membrane potential which has been interrogated in any kind of routine manner. Lately, by utilising appropriate probes (eg. as in the present chapter, di-8-ANEPPS and FPE), the spatial heterogeneity of the other membrane potentials may also be visualised.

In our laboratories we developed the use of di-8-ANEPPS to determine how the membrane dipole potential may be utilised to reveal macro-molecular interactions with membranes. By illuminating spatial variations of this parameter and particularly how it varies in membrane rafts it led us to propose that the value of ϕ_d is quite different in membrane rafts as compared to fluid mosaic membranes. Figure 5.6 illustrates this striking heterogeneity of the ϕ_d about the cell surface with measures currently being taken to correlate this or co-locate such signals that are emanating from the raft microdomains within membranes. The latter is a significant issue as there remains much confusion as the *in vivo* existence of membrane rafts. This facilitates the detections of interactions (ie. within single cells) that lead to changes of the value of ϕ_d may be revealed.

Similar approaches utilising such indicators as FPE to visualise the membrane surface potential ϕ_s are also routinely employed in our laboratories. By correlating the change of the fluorescence and hence surface potential with the addition of net electric charges from the macromolecule that becomes bound, it is possible to quantitate on the basis of the Poisson-Boltzmann equation above, the number of molecules that become bound. This allows us for example to determine localised molecular interactions on the membrane surface. Figure 5.7 illustrates a typical image for the binding of human serum albumin (HSA) to a single fibroblast cell.

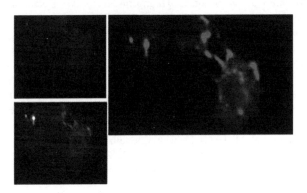

Figure 5.7 Electrostatic map of the cell surface

Summary

This chapter outlines the basic doctrines of the structure of biological membranes, including the involvement of microdomains in membranes known as rafts. Some of the basic functions of membranes concerned with electrical activity are outlined in more depth with particular emphasis in the physical origins and formal (mathematical) expressions of these 'membrane potentials'. The latter is outlined by taking account of the differing nomenclatures utilised by biological and chemical scientists. Their roles in many different biological processes are outlined and several outstanding questions identified. This is further developed by describing several techniques for the measurement of these potentials with a view to their experimental implementation in biological systems. Finally, some aspects of the spatial imaging of these ubiquitous membrane properties are described in living cells.

Acknowledgements

I am grateful to Angus Bain (London), Peter Bond (Oxford), Martin Karplus (Boston) and Mark Sansom (Oxford) and Mike Somekh (Nottingham) for helpful discussion as well as to my colleagues within the Cell Biophysics Group. The Cell Biophysics Group is funded by research grants from the Wellcome Trust, Research Councils UK (RCUK) and the EPSRC, it is a pleasure to acknowledge their valuable support as well as the general support from the University of Nottingham.

Further reading

Bockris JO'M, Reddy AK (1970). Modern electrochemistry. MacDonald, London.

Bunting JR, Phan TV, Kamali E, Dowben RM (1989). Fluorescent cationic probes of mitochondria. Metrics and mechanism of interaction. *Biophys J.* **56**: 979–93.

Cladera J, O'Shea P (1998). Intramembrane molecular dipoles affect the membrane insertion and folding of a model amphiphilic peptide. *Biophys J.* **74**: 2434–2442.

Lagerholm BC, Weinreb GE, Jacobson K, Thompson NL (2005). Detecting microdomains in intact cell membranes. *Annu. Rev. Phys. Chem.* **56**: 309–36.

Nicholls D, Ferguson S (1996). Bioenergetics, 2nd Edition. John Wiley & Son, Ltd., Chichester.

O'Shea P (1988). Physical fields and cellular organisation. *Experientia* **44**: 684–694.

O'Shea P (2003). Intermolecular interactions with/within cell membranes and the trinity of membrane potentials: kinetics and imaging. *Biochem Soc Trans* **31**: 990–996.

O'Shea P (2004). Membrane Potentials; measurement, occurrence & roles in cellular function pp 23–59 in *Bioelectrochemistry of Membranes* (D. Walz *et al.* Eds) Birkhauser Verlag, Switzerland.

Ross E, Bedlack RS, Loew LM (1994). Dual-wavelength ratiometric fluorescence measurement of the membrane dipole potential. *Biophys J.* **67**: 208–16.

Wall JS, Ayoub F, O'Shea P (1995). The interactions of macromolecules with the mammalian cell surface. *J Cell Sci* **108**: 2673–2682.

6

Identification and quantification of lipids using mass spectrometry

Trevor R. Pettitt and **Michael J. O. Wakelam**

CR-UK Institute for Cancer Studies, Birmingham University, Birmingham B15 2TT, UK

6.1 Introduction

Glycerophospholipids and sphingophospholipids are major structural constituents of cellular membranes. However, they are also important substrates for signal generating enzymes and indeed in a number of cases have been found to function as intra- or extra-cellular messengers. Thus it is of great importance to determine the levels of individual lipids in cells and tissues and to define changes in response to physiological and pathological stimuli. Examination of the literature will show that, until recently, many such analyses have relied upon differences in radiolabelling, in particular making use of [^{32}P] for labelling the lipid headgroup and [^{3}H] for labelling the fatty acid constituents. Owing to the problems associated with achieving true isotopic equilibrium when radiolabelling (this can often take significantly longer than the 16–24 h many people assume), a great deal of effort has been put into developing nonradioactive methods that can determine changes in the levels of signalling lipids in particular. Much of this has involved thin-layer chromatography and/or assays of enzyme activities. Whilst these methodologies have undoubtedly provided much data of importance, there are severe limitations to these approaches, not least because a single lipid class comprises many distinct molecular species defined by their radyl (acyl, alkyl, alkenyl) compositions. Sensitivity has also been a big issue since most TLC detection methodologies cannot detect much below 1 μg of any particular lipid without the use of a radiolabel.

Chemical Biology   Edited by Banafshé Larijani, Colin. A. Rosser and Rudiger Woscholski
© 2006 John Wiley & Sons, Ltd

Identification of the fatty acid composition of glycerolipids became possible by making use of gas chromatography (GC), but this normally requires hydrolysis and derivitization of the lipids to make them volatile prior to analysis. Coupling GC to a mass spectrometer (GC-MS) increased the structural information obtainable, but this methodology was still unable to analyse most intact lipid molecules, particularly the nonvolatile phospholipids which rapidly pyrolyse at the temperatures used for GC. As with the analysis of proteins and peptides, the major advances in lipid analysis have come about through recent developments in mass spectrometry together with replacement of TLC by improvements in separation by HPLC. Together, these two technologies substantially increase the resolution and sensitivity for lipid analyses.

6.2   Lipid analysis by mass spectrometry

In addition to the GC-coupled MS methods discussed above (electron impact and chemical ionization), fast atom bombardment (FAB), particle beam and thermospray mass spectrometry have been used in the analysis of phospholipids, lysolipids, diglycerides and fatty acids (Matsubara and Hayashi, 1991; Aberth *et al.*, 1982; Gross, 1984). More recently, atmospheric pressure chemical ionization (APCI; Byrdwell, 2001) and matrix-assisted laser desorption ionization (MALDI) has been used for some analyses. However, it was the development of electrospray ionization mass spectrometry (ESI-MS), together with tandem mass spectrometry (MS/MS) that made the precise identification and quantitation of global lipid species a feasible aim. The essential problem with most previous techniques, such as FAB, is that they are relatively harsh, inducing extensive fragmentation of the lipids. Whilst this is useful for determining exact structures based on characteristic fragmentation patterns, it usually requires an essentially pure compound; otherwise, as in unresolved lipid samples composed of many closely related structures, the complex fragmentation patterns become largely uninterpretable.

ESI is a 'soft', atmospheric pressure ionization technique that works by mechanisms that are not fully understood and still open to debate. The sample dissolved in an appropriate solvent is passed through a spray needle maintained at high voltage (1–6 kV), creating a spray of charged droplets. A nebulizing or sheath gas, usually nitrogen, is often applied, particularly at higher solvent flow rates, to assist with droplet formation and subsequent solvent evaporation. As this desolvation proceeds, usually aided by some heating, the droplets get smaller, increasing the number of ions at the droplet surface. When the surface charge density exceeds a critical threshold (where the Coulombic repulsion of these ions exceeds droplet surface tension), droplets will fission into smaller daughters and the process repeats. Ultimately, as the repulsive forces get too great and/or further fission is no longer possible, ion ejection occurs and these pass into the ion optics of the MS where they can be analysed. Incomplete desolvation can lead to charge-carrying solvent clusters entering the detector and contribute to random background noise. Depending on the polarity applied to the spray needle, positive or negative ions can be analysed. Anionic

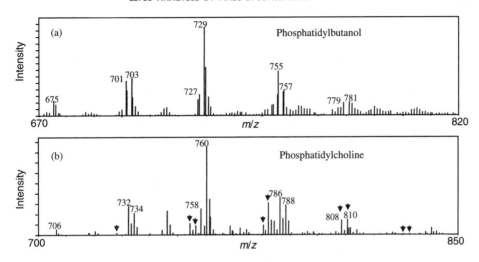

Figure 6.1 Examples of ESI mass spectra of phospholipids. (a) Negative ESI mass spectra of mammalian PtdBtOH [M-H]⁻. (b) Positive ESI mass spectra of mammalian PtdCho [M + H]⁺. Arrowed peaks indicate sodiated adducts [M + Na]⁺ of PtdCho

phospholipids {phosphatidylalcohols, phosphatidic acid (PtdOH), phosphatidylglycerol (PtdGly), phosphatidylinositol (PtdIns), phosphatidylserine (PtdSer), cardiolipin} give best ionization in negative mode, which normally creates a pseudomolecular ion following the loss of a proton, [M-H]⁻ [Figure 6.1(a)]. Negative ionization also tends to give the cleanest mass spectra with least adduct formation, thus making interpretation easier. Phosphatidylethanolamine (PtdEtn) gives reasonable ionization in both positive and negative modes whilst phosphatidylcholine (PtdCho) and sphingomyelin (SM) give poor negative ionization, usually through loss of a methyl group [M-15]⁻, but good positive ionisation [M + H]⁺ [Figure 6.1(b)]. However their spectra are complicated by the presence of sodium [M + Na]⁺ and potassium [M + K]⁺ adducts formed through the abstraction of the metal from surfaces and/or solvents. These adducts probably cannot be totally prevented, although the inclusion of modifiers such as LiOH can shift most of the ions to an alternative adduct, in this case [M + Li]⁺ (Han and Gross, 2003). The use of solvent modifiers can also allow ion generation from molecules that lack normally ionizable sites. For instance, ammonium formate and ammonium acetate will allow positive ionization of diradylglycerol (DRG; forming [M + Na]⁺), whilst triradylglycerols can be ionized in the presence of LiOH, both molecules which otherwise show little or no ionization.

One drawback to ESI-MS is that ideally all solvent modifiers should be volatile, otherwise salt build-up on the MS orifice will lead to signal degradation unless regularly cleaned away. Thus the use of phosphate and similar nonvolatile buffers is best avoided. The exact ion profiles generated can vary somewhat between instruments for reasons which are not entirely clear, although this probably relates to differences in probe and interface designs.

Another potential limitation of ESI-MS is that under most circumstances it is probably best viewed as semiquantitative even after correction for natural isotope, distributions. This is particularly relevant in relation to lipid analyses where there are only a small number of usable standards (heavy isotopes, e.g. deuterated analogues, or nonphysiological radyl structures, e.g. dilauroyl, 12:0/12:0, or dieicosanoyl, 20:0/ 20:0, species; in this lipid shorthand the number before the colon represents the number of carbons in the fatty acid and the number after the colon represents the number of double bonds in the fatty acid) are commercially available. Equimolar amounts of different lipid classes will show different ion intensities, as will equimolar amounts of a given lipid class, but with different radyl compositions. For instance, Koivusalo *et al.* (2001) demonstrated that instrument response decreased with increasing acyl chain length and increased with double bond number, although both these effects declined at lower total lipid concentrations. DeLong *et al.* (2001) have shown linearity between infused total lipid concentration and ion intensity for PtdCho and PtdEtn below 2 μM, but steadily lost linearity at higher concentrations. A number of groups have developed algorithms attempting to correct for these response differences (e.g. Zacarias *et al.*, 2002); however these are probably only applicable to the instrument and conditions used in that work. Since relative abundances, in lipid species are usually more important than absolute abundances, these problems are not necessarily of significant concern to most analyses.

The sensitivity of MS analysis has been a boon in facilitating the determination of lipid compositions; however this very sensitivity has also been a hindrance because of the complexity of a physiological lipid mixture. This has led to the adoption of two complimentary approaches in the analysis of cellular lipids using MS. In the first, advances in separation methodologies particularly in HPLC, have been used to reduce the complexity of the sample entering the MS, whilst the second has been the adoption of ESI-MS/MS tandem mass spectrometry. The advantages of the two methods are considered below, although the benefits of each suggest that both will remain in use and that an 'ideal' method will probably adopt aspects of both in complete lipid analyses.

6.2.1   HPLC-ESI-MS

Use of the HPLC-ESI-MS methodology with single quadrupole instruments has been primarily in the determination of particular classes of lipids rather than in attempting a more 'global' lipidomic analysis. This MS identification methodology has been used in determining changes in lipids thought to be involved in a particular cellular process, or alternatively to determine changes in lipids which can act as markers for such events. Examples of both of these concepts are reflected in analysing the activity of phospholipase D (PLD) in cells. PLD catalyses the hydrolysis of PtdCho, thereby generating the signalling lipid PtdOH and choline as products. In the presence of a short chain primary aliphatic alcohol, e.g. butan-1-ol, PLD preferentially catalyses the generation of a phosphatidylalcohol, e.g. phosphatidylbutanol (PtdBtOH). Optimization of the separation conditions for total cellular lipid extracts upon a silica

HPLC column using a chloroform–methanol–water–ammonia gradient prior to ESI-MS made it possible to resolve most phospholipid classes. Additionally, a partial resolution of component molecular species is also possible using this procedure, since the longer chain species elute slightly earlier than the shorter species whilst the more unsaturated species are eluted ahead of the corresponding more saturated forms. Using this methodology, with the inclusion of appropriate internal standards, it has been possible to determine growth factor-stimulated PLD activity; the availability of such standards also permits a certain level of quantitation. This analysis has shown increased agonist-stimulated PtdOH formation in the absence of butan-1-ol, a decrease in its formation in the presence of the alcohol and under such conditions an increase in PtdBtOH formation (Pettitt *et al.*, 2001). This analysis also permitted the determination of the lipid species in each class of interest (Figure 6.2). This has been particularly useful in demonstrating that the PtdOH species generated by PLD activation are predominantly mono- and diunsaturated in mammalian, yeast and *Dictyostelium* cells, whilst those generated as a consequence of phospholipase C-catalysed phosphatidylinositol 4,5-bisphosphate hydrolysis and subsequent diacyl-glycerol kinase-catalysed diacylglycerol phosphorylation are predominantly polyun-saturated, thereby demonstrating distinction between 'signalling' and 'nonsignalling' PtdOH as being due to the extent of unsaturation.

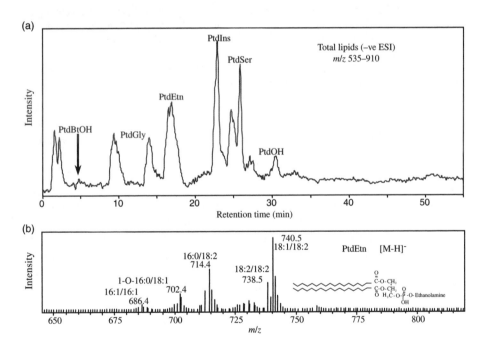

Figure 6.2 (a) Total ion current (TIC) chromatogram of *Dictyostelium discoideum* total lipid extract separated on a silica HPLC column using a gradient of 100 % chloroform–methanol–water–ammonium hydroxide (90:9.5:0.5:0.32), changing to 100 % chloroform–methanol–water–ammonium hydroxide (50:48:2:0.32) over 45 min with ESI-MS detection in negative mode and a scan range of m/z 535–910. (b) Mass spectra extracted from the PtdEtn peak

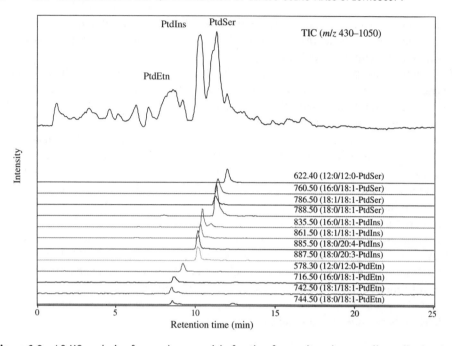

Figure 6.3   LC-MS analysis of a membrane vesicle fraction from cultured mammalian cells showing the negative ESI-MS TIC and extracted chromatograms for the major PtdEtn, PtdIns and PtdSer molecular species, together with the 12:0/12:0-PtdEtn and PtdSer internal standards. Lipids separated on a silica HPLC column using a gradient of 100 % chloroform–methanol–water (90:9.5:0.5) containing 5 mM ethylamine changing to 100 % acetonitrile–chloroform–methanol–water (30:30:35:5) containing 10 mM ethylamine over 20 min

---

HPLC-ESI-MS methods have been able to determine much of the total phospholipid composition of complete extracts from *Dictyostelium*, yeast and *Drosophila* samples as well as various mammalian whole cell and subcellular fractions (Figure 6.3). This has included the identification of sn1-alkyl, 2-acyl- and sn1-alkenyl, 2-acyl-phospholipids in addition to the sn1,2-diacyl-structures. Despite these successful analyses and the potential for further advances in separation methodology, the use of ESI-MS is limited by its inability to produce controlled, diagnostic sample fragmention, which is needed if the lipid class is to be unambiguously assigned and the exact radyl composition is to be determined rather than just indicating a group of isobaric species, e.g. 18:0/20:4, 18:1/20:3 and 16:0/22:4, or alternatively C38:4.

Concerns have been raised about the stability of certain lipids on HPLC, although these issues are probably largely dependent on column chemistries and/or solvent compositions and, as such, should be solvable. Losses of material due to irreversible column binding may be a problem when working with very low amounts of lipid

Table 6.1 Characteristics of ESI-MS

Advantages
High sensitivity
Usually only need small amounts of material
Can provide full structural identification
Derivatization not normally required
Disadvantages
Expensive equipment
Often only semiquantitative
Not all structures ionize well
High sample or contaminant concentrations can cause
   ion suppression
Solvents and modifiers should be volatile

and here direct sample infusion into the MS may be the most appropriate choice, particularly when only a few defined lipid structures are of interest. Nevertheless characterization of HPLC methodologies and the use of appropriate standards have permitted this column method to become that of choice. The advantages and disadvantages of ESI-MS are outlined in Table 6.1, whilst the main detectable ions are listed in Table 6.2.

6.2.2 Tandem mass spectrometry

The increasing use of triple-quadrupole and ion-trap mass spectrometers has significantly enhanced the ability to perform structural identification of multiple species in a single analysis. An increasing number of reports have made use of direct injection of lipid mixtures with no prior chromatographic lipid separation.

Table 6.2 Major ions

Lipid	Detected as
PtdOH	$[M-H]^-$
PtdBtOH	$[M-H]^-$
PtdCho	$[M+H]^+$, $[M+Na]^+$
PtdEtn	$[M-H]^-$, $[M+H]^+$
PtdGly	$[M-H]^-$
PtdIns	$[M-H]^-$
PtdInsP	$[M-H]^-$, $[M-H]^{2-}$
PtdInsP$_2$	$[M-H]^-$, $[M-H]^{2-}$
PtdSer	$[M-H]^-$
DRG	$[M+Na]^+$
SM	$[M+H]^+$, $[M+Na]^+$
LysoPtdOH	$[M-H]^-$
LysoPtdCho	$[M+H]^+$
FFA	$[M-H]^-$

Whilst there will be overlap of different species, the use of collision-induced dissociation (CID) of individual species generates fragment ions (MS^2) which can be used to identify the lipid class and radyl composition (Pulfer and Murphy, 2003). Studies of the major fragmentation pathways of phospholipids in a triple quadrupole mass spectrometer have suggested that there is nucleophilic attack on the sn-1 and sn-2 glycerol carbons by the phosphate oxygen. This leads to charge transfer forming carboxylate anions associated with the neutral loss of a five- or six-membered ring of a cyclo-lysophospholipid (see Larsen *et al.*, 2001). The use of appropriate software thus permits determination of the parent structures.

Unfortunately, not all molecular species are detected with equal efficiency by MS. The detection of an individual phospholipids is dependent both upon radyl chain saturation and length with a greater intensity of response being determined for more unsaturated species, but a lower response being found as the chain length increased. In addition the detection of the different lipid classes is affected by solvent since this can alter the nature of the adduct formation and thus simplify or make more complex the spectra. The CID fragmentation of lipids in most cases permits the definitive identification of individual species.

A major problem associated with this method of analysis, aside from the difficulties in deconvolving extremely complex spectra when analysing a total lipid extract, is the inability to distinguish between structural isomers such as the different mono- and bisphosphorylated forms of phosphatidylinositol; these show the same pseudomolecular ions and undergo identical fragmentations but are functionally distinct; therefore there remains a need to develop methods to unambiguously identify each of these species.

Despite these caveats, Schneiter *et al.* (1999) employed nano-ESI-MS/MS to determine differences between lipid species compositions of organelle membranes prepared from *S. cerevisiae* cells. These studies showed differences in degree of saturation between phospholipids with the highest level being found in PtdIns and the lowest in PtdCho; intriguingly the polyphosphoinositides are far more unsaturated in mammalian cells. The analysis did demonstrate plasma membrane enrichment of saturated PtdSer and PtdEtn species, although in general the species profile was rather similar between the different membranes. Analysis of appropriate mutants showed that the plasma membrane accumulation of monounsaturated rather than diunsaturated PtdSer was a consequence of membrane-specific acyl chain remodelling. This analysis highlighted the potential of lipidomics to define metabolically distinct membranes and thus the possibility of determining changes in a particular membrane rather than in a whole cell. This type of analysis will be of particular importance in examining receptor-stimulated signalling events involving lipid-modifying enzymes such as phospholipases and lipid kinases. It will also be important in defining the specific lipid composition of functionally distinct membranes and domains, e.g. secretory vesicles, endosomes, plasma membranes, lipid rafts (see e.g. Figure 6.3).

6.3 Conclusion

ESI-MS/MS has been successfully adopted by groups such as Schneiter *et al.* (1999) to determine much of the lipid composition of membranes. By combining the best aspects of HPLC and tandem mass spectrometry it should be possible to extend this in order to achieve global lipidomics. In this a total lipid extract will be analysed for all its component lipids and the data compared with appropriate control samples and displayed in a suitably graphic style so that any changes, either positive or negative, are easily viewed. This is something that is being actively worked on by several groups (Forrester *et al.*, 2004; Han *et al.*, 2004) and should give a good chance of finding that proverbial 'needle in a haystack'.

Acknowledgements

Work in this laboratory is funded by the Wellcome Trust, Cancer Research UK and the MRC.

Further reading

Aberth, W., Straub, K.M. and Burlingame, A.L. (1982). *Anal. Chem.* **54**: 2029–2034.

DeLong, C.J., Baker, P.R.S., Samuel, M., Cui, Z. and Thomas, M.J. (2001). *J. Lipid Res.* **42**: 1959–1968.

Forrester, J.S., Milne, S.B., Ivanova, P.T. and Brown, H.A. (2004). *Mol. Pharmac.* **65**: 813–821.

Gross, R.W. (1984). *Biochemistry* **23**: 158–165.

Han, X. and Gross, R.W. (2003). *J. Lipid Res.* **44**: 1071–1079.

Han, X.L., Yang, J.Y., Cheng, H., Ye, H.P. and Gross, R.W. (2004). *Anal. Biochem.* **330**: 317–331.

Koivusalo, M., Haimi, P., Heikinheimo, L., Kostiainen, R. and Somerharju, P. (2001). *J. Lipid Res.* **42**: 663–672.

Larsen, A., Uran, S., Jacobsen, P.B. and Skotland, T. (2001). *Rapid Commun. Mass Spectrom.* **15**: 2393–2398.

Matsubara, T. and Hayashi, A. (1991). *Prog. Lipid Res.* **30**: 301–322.

Pettitt, T.R., McDermott, M., Saqib, K.M., Shimwell, N. and Wakelam, M.J.O. (2001) *Biochem. J.* **360**: 707–715.

Pulfer, M. and Murphy, R.C. (2003). *Mass Spectrom. Rev.* **22**: 332–364.

Schneiter, R., Brugger, B., Sandhoff, R., Zellnig, G., Leber, A., Lampl, M., Athenstaedt, K., Hrastnik, C., Eder, S., Daum, G., Paltauf, F., Wieland, F.T. and Kohlwein, S.D. (1999). *J. Cell Biol.* **146**: 741–754.

Zacarias, A., Bolanowski, D. and Bhatnagar, A. (2002). *Anal. Biochem.* **308**: 152–159.

7

Liquid-state NMR

Charlie Dickinson

The University of Massachusetts, Amherst, MA, USA

7.1 Introduction

Nuclear magnetic resonance (NMR) is a pre-eminent technique for molecular-level understanding because it exquisitely displays differences in and connections between *chemical* environments of atoms in a molecule. NMR began in 1946 as a technique for physicists concerned only with nuclear properties, but quickly evolved. By NMR chemists could readily 'see' a very useful representation of molecular structure. In the past 25 years, with vastly improved sensitivity and sophistication and imaging capability, many areas of biological sciences have taken advantage of the power of NMR to observe selected components of organisms, both *in vivo* and *in vitro*. This chapter will show a few examples of such biological applications in order to illustrate the potential of NMR to supply salient details in almost any biological research area. Chemists and biochemists have emphasized NMR as a molecular structural tool and indeed complete three-dimensional structures of proteins up to and over 20 kDa molecular weight have been determined in solution. This chapter presents only a representative structural study of a peptide, but the reader should examine the Further Reading (especially the book by Evans, 1995) to grasp the structural triumphs of solution NMR in obtaining detailed structures of proteins nucleic acids and polysaccharides.

There are myriad biological applications of NMR to almost every tissue and fluid of animals and plants already in the literature. This chapter is merely a bridgehead into a vast field.

Chemical Biology   Edited by Banafshé Larijani, Colin. A. Rosser and Rudiger Woscholski
© 2006 John Wiley & Sons, Ltd

Box 1   Advantages and limitations of NMR

Advantages of NMR

- Specific molecules can be quickly identified and quantified even in complex mixtures.

- Ease and speed of measurement often allows real-time kinetics to be followed.

- Complete three-dimensional structures can be determined in solution even for proteins as big as 20 kDa.

Limitations of NMR

- Relative to mass spectrometry and photon methods, NMR is a weak technique for other than proton observation, but high magnetic fields (900 MHz) allow detection of tens of microgram quantities in a matter of seconds.

- Working within the constraints of a superconducting magnet bore sometimes restricts experimental design for *in vivo* systems.

- Viscous or gel samples require special probes with spinning at 2kHz to achieve good resolution.

7.2   How NMR works: the basics

In a strong magnetic field (the arrow labeled $\mathbf{B_o}$ in Figure 7.1) all magnetic nuclei of atoms 'precess', or spin, like a gyroscope around the z-axis at a frequency governed by the nature of the nucleus and its chemical environment. The precession frequency is proportional to the magnetic field strength. Magnets are referred to by the frequency at which protons precess in their field, for example, 600 MHz would be a typical 'magnetic field' for a modern spectrometer, but magnets range up to 900 MHz. In a 600 MHz field, carbon-13 nuclei will precess at 150 MHz, and phosphorus-31 at 242 MHz. Only some nuclei have magnetic moments; for example, carbon-12 or magnesium-24 are not NMR active as they have no magnetic moment, yet carbon-13 (1.1 per cent natural abundance) and magnesium-25 (10.1 per cent natural abundance) are NMR active. *Almost any element in the periodic table has at least one magnetic isotope and there is NMR experience with almost every element.* The energy of a magnetic moment of a nucleus interacting with a magnetic field is small relative to the ambient thermal agitation energy and this makes NMR a relatively insensitive observational technique, especially when compared with mass spectrometry or photon detection techniques. Nevertheless, the strong magnetic field

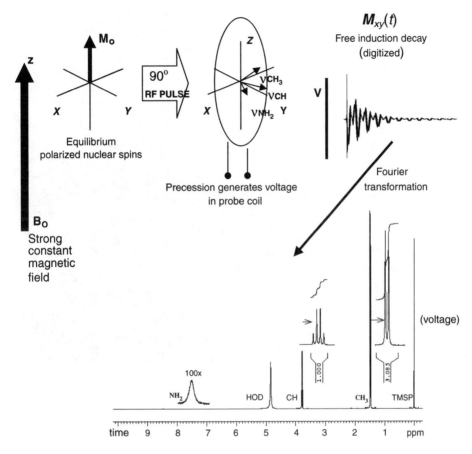

Figure 7.1 Schematic of the detection and processing of nuclear magnetic resonance signals. The spectrum shown is of alanine in D_2O with details as explained in the text

polarizes the nuclei such that more of them are pointing with the field than against it; there is a net magnetization vector, M_o, pointing with the field as illustrated in Figure 7.1. If there is a variety of chemical environments, e.g. CH_3, NH_2 and CH protons on the α-amino acid alanine, each will contribute a precession frequency component to M_o. At equilibrium M_o is unobservable because the sample is a large ensemble of nuclei; instead of a single nucleus precessing around the z-axis, the representation is a cone of vectors spread evenly around the cone and their xy-plane values cancel out. A pulse of electromagnetic radiation in the form of a radio frequency wave at on near the precession frequency causes the magnetization vector, M_o, to bend away from the strong magnetic field. Precise timing of the length of the *radiofrequency pulse* can cause the M_o vector to tip 90° and each component of M_o becomes R_{xy} (t) and continues to process at its characteristic frequency in the x,y-plane. Each precessing magnetic component induces a voltage in the detection coil *at its characteristic frequency*, ν_i. This induced voltage pattern is called the *free induction decay* (FID), which is recorded in a digitized form (Figure 7.1). The

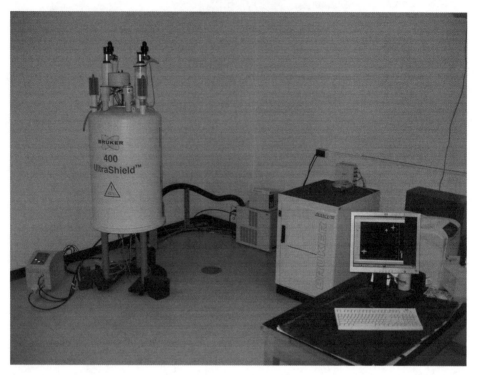

Figure 7.2 A modern NMR spectrometer showing from left to right a 400 MHz superconducting magnet, electronics console including an acquisition computer and radiofrequency amplifiers, and keyboard and monitor for the PC which inputs parameters and commands to, and extracts data from, the spectrometer

pulse experiment can be repeated n-fold times, resulting in an $n^{1/2}$ increase in the signal-to-noise ratio. This digitized intensity vs time pattern (the superposition of all the contributing frequencies) is *Fourier transformed* (FT) via a particularly efficient computer algorithm to result in a pattern of intensity vs frequency, which is a normal one-dimensional NMR spectrum, as shown in Figure 7.1.

As shown in Figure 7.2, an NMR spectrometer consists of a strong magnet, a 'radio station' for sending pulses and a radio frequency receiver to detect the FID, and a computer to control acquisition, record and transform the data. Frequencies for a given nucleus are referenced to a standard, for example the proton peak of tetramethylsilane is assigned 0 frequency and all proton peaks are related to that position in terms of *chemical shifts* in parts per million (ppm). Chemical shifts place resonances of different types in characteristic regions; for example aromatic protons appear at 6–7 ppm, methyl protons at 0.5–1.5 ppm, amide protons at 7.5–9 ppm, etc. The representation of the acquisition of a proton.

The NMR spectrum of alanine shown in Figure 7.1 illustrates that the different types of protons are well resolved and that one can 'count' the relative number of protons for each chemically distinct species by integrating the peaks areas: the ratio of the CH_3 peak to the CH peak is very nearly the expected 3.0. The NH_2 peak of the alanine is very much weakened because these protons are labile, that is, the protons exchange with the deuterons of the D_2O. The NH_2 resonance is also broadened because the exchange rate shortens the lifetime of the proton states. If the alanine were dissolved in an aprotic solvent so that the NH_2 remained protonated, its resonance would be narrower and the CH_3 and CH resonances would show further splittings. Note also that a residual resonance appears from the residual H_2O in the 99.9 per cent D_2O solvent. Also note that there is a peak at 0 ppm from 3-(trimethylsilyl)propane sulfonic acid sodium salt (TMSP), added as a standardization reference for aqueous solutions. Horizontal expansion of the peaks at 1.37 (CH_3) and 3.66 (CH) ppm reveals spin–spin splittings of the peaks which yield further details of molecular structure. These splittings, characterized by J, the *spin–spin coupling constant*, depend on the *through-bond* coupling of one spin to another via the electrons. Three-bond proton–proton coupling constants as seen here are typically 5–10 Hz and reflect in a complex way the nature of the bonds. In the alanine spectrum the CH_3 protons are split into a doublet because they can magnetically 'see' the CH proton, which can be either with or against $\mathbf{B_o}$, shifting the line by $\pm J/2$. Similarly the spins of the three CH_3 protons can be ordered in four distinct ways in the field and these split the CH proton into a quartet of peaks centred at the CH chemical shift.

The nuclear spins of an NMR sample can also be subjected to a choreographed series of pulses which capitalize on the correlation between spins. For example, choosing a pulse program to detect protons that are spin–spin coupled results in a spectrum displayed as a *two-dimensional (2D) NMR* plot referred to as correlation spectroscopy (COSY). The peptide example below shows several other types of proton–proton 2D NMR. If a pulse program selecting the correlation between protons and the carbons (or phosphorus or any other nucleus) to which they are coupled is chosen, one obtains a heteronuclear multiple quantum correlation (HMQC). (See the phospholipid example below.) Correlation among all spins that are chained together by J couplings is called a TOCSY (total correlation spectroscopy, see the peptide example below). For correlations through space there is NOESY (nuclear overhauser enhancement spectroscopy, see the peptide example). There are a vast number of two-, three- and four-dimensional pulse programs available in the current NMR arsenal for detecting specific correlations relevant to discerning bondings or proximities of nuclei: see Berger and Braun (2004).

The dynamics of the nuclear spins, that is, how fast $\mathbf{M_o}$ or $\mathbf{M_{xy}}$ changes after the pulse, carries a lot of information about the mobility of the molecular environment of the species observed. These dynamics are measured by the rate at which $\mathbf{M_o}$ decays back to equilibrium along the large magnetic field by the *spin–lattice relaxation* (rate constant $1/T_1$) mechanism or the rate at which $\mathbf{M_{xy}}$ fans out to unobservability, the *spin–spin relaxation* (rate constant $1/T_2$) mechanism. The decay rate of the FID in Figure 7.1 is $1/T_2$. These relaxation times govern how fast an NMR experiment can be repeated. They are a well-understood complex function of the rate at which the molecule can tumble, which is governed by the viscosity of the environment, size of the molecular species or its aggregates, and many other local motion factors. In biological systems T_2 and/or T_1 decays of magnetization are often multicomponent and these curves have been quantitatively analysed to describe various microanatomical components in peripheral nerves and mobility and mechanisms of crystallization of starches, and a host of other heterogeneous processes.

As a practical matter only a few milligrams in *ca.* 0.5 ml of solvent are required for a normal 1D proton NMR, as shown in Figure 7.1, but much can be done with a lot less. For ^{13}C-NMR at natural abundance, more sample is required. There is a tradeoff between the amount of sample and the time on the spectrometer. Often a single scan is sufficient for a proton NMR so that real-time kinetics can be observed. Standard sample tubes are 5 mm in diameter with an *in vitro* sample filling the bottom 4 cm of an 18 cm tube, but there is a wide variety of other tubes and special probes for selective experiments. Also, as a practical matter samples are generally dissolved in deuteriated solvents in order to both reduce the strong solvent proton signal and to give a deuterium NMR signal which is used to stabilize the potential drift of the large magnetic field. This latter is the 'lock' signal for field-frequency lock. The capped sample tube grasped by a small holder is pneumatically lowered into the coil of the probe through the room-temperature centre tube of the superconducting magnet.

Because of the complexity of many modern NMR experiments and the customizing necessary to adapt sometimes complicated NMR techniques to specific analytical cases, most major research centres hire staff NMR specialists to collaborate in getting the NMR experiments of users to be accurate and productive. In this era the beginning biological researcher should not feel the necessity of mastering all of NMR in order to take advantage of its power.

7.3   Some NMR applications in biology

7.3.1   *In vivo* cell metabolism made 'visible' by NMR

It is rather simple in most cases to use NMR to monitor certain processes *in vivo* in whole cells (Rager *et al.*, 2000). In this example the researchers were curious about the specific metabolic path by which the pathogenic bacterium *Plesioimonas shigelloides* can use glucose but not mannose. The researchers simply grew cells to the late-exponential stage, centrifuged them down, washed nutrients out with phosphate buffer, centrifuged again and resuspended in buffer in an NMR tube at 30 °C. For the time-course study, they injected glucose at time zero and monitored the change in the ^{13}C -NMR spectrum as a function of time, the *in vivo* process occurring directly in the NMR tube while in the magnet.

^{13}C NMR signals are usually about 10^{-4} weaker than proton NMR signals because only 1.1 per cent of the carbons are magnetic and because the resonant frequency of ^{13}C is one-quarter that of protons. Nonetheless, ^{13}C NMR can be routinely observed in a reasonable time *in vitro* for low-molecular-weight samples available in tens of milligram quantities. However, if one obtains the ^{13}C-NMR spectrum of whole cells, the signal will be weaker because concentrations are low, the lines are broadened and there will also be a vast number of different types of carbon observed, resulting in a complex and overlapped spectrum. One way to greatly enhance the ^{13}C-NMR signal and to limit the species observed to the desired ones is to purchase materials which have been enriched even to 100 per cent [13]C. Typically the cost is US \$200/g for off-the-shelf ^{13}C-enriched materials, and there are commercial laboratories which perform custom synthesis.

In this case the researchers purchased carbon-13 enriched [1-^{13}C]glucose and [6-^{13}C]mannose and at time zero mixed each with cell suspensions of the bacteria. The sample was placed directly in an NMR probe and the signals accumulated over 256 repeated scans and the resulting FID were Fourier-transformed. The time series spectra in Figure 7.3 show that the glucose was consumed and converted to [3-^{13}C]acetate (21.1 ppm), [2-^{13}C]-ethanol (17.8 ppm) and lower concentration products not yet observable in the short time span shown. (Note also that the two optically active D- and L-anomers of [1-^{13}C]glucose are present and give separate ^{13}C resonances at 96.8 and 93 ppm, respectively, yet both are consumed at the same rate.) Mannose, on the other hand, was slowly converted only to (6-^{13}C)mannose-6-phosphate (M6P at 64.4 ppm) with time. (Phosphorylating the 6 position of the mannose shifts the ^{13}C peak of carbon 6 to the left or 'downfield' by about 3 ppm, showing clearly how chemical shift reflects chemical changes.) This result shows the novel result that mannose is transported into the cell but is not metabolically consumed past the 6-phospate product. M6P becomes toxic to the cell as the concentration grows. Other assay data, including ^{31}P-NMR data (not shown), were used to show that the mannose metabolic bottleneck occurs because glucose-6-phosphate (G6P) isomerase isomerizes the G6P moiety but not the M6P into the metabolically consumable fructose-6 phosphate.

This technique of simply examining the chemically selected components of whole cells *in vivo* via isotopic enrichment has many, many powerful applications in the biological literature. Furthermore, simple one-dimensional NMR has also been used *in vitro* to great advantage, even without isotopic enrichment, to analyse content and changes with pathologies on almost all biofluids, such as blood, urine, bile, synovial fluid, tears and saliva. The capacity to examine *in vivo* cellular tissue or fluids has been extended by magnetic resonance imaging spectroscopy (MRIS). MRIS combines the spatial specificity of imaging (not addressed here) with the type of simple NMR spectroscopy given in this example. One can for example examine the molecular changes in phosphorus metabolism in selected small volumes of the brain in a living person.

7.3.2 *In vivo* phosphorus and nitrogen metabolism

The specific fate and effects of various chemicals – helpful or harmful – on plants is of great importance in understanding plant metabolism and pathology (Gerendas and Ratcliffe, 2000). The internal effects of ammonia/um nutrient can be monitored *in vivo* in maize roots via NMR: NH_3 concentration can be monitored by ^{14}N-NMR and pH, and the phosphorylated species concentration can be measured by ^{31}P-NMR and synthesis of carboxylate followed by ^{13}C -NMR. The specific fate of the NH_3 can also be followed by ^{15}N-NMR with isotopically enriched feed species.

(a)

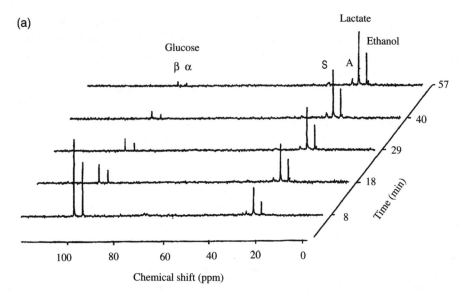

(b)

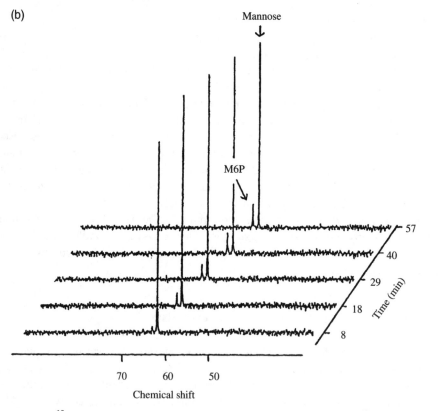

Figure 7.3 ^{13}C NMR spectra of *in vivo* kinetics of the consumption of (a) [1-^{13}C]-glucose and (b) [6-^{13}C]-mannose as followed using ^{13}C-NMR by *P. shigelloides*. Details are given in the text. [Reproduced from Rager *et al.* (2000) *Eur. J. Biochem* **267**: 5136–5141]

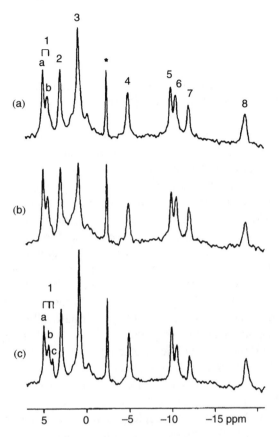

Figure 7.4   Phosphate species observed in maize root (a) before (b) 90 min after addition of 20 mM ammonium and (c) after 3 h of recovery. Signals 2 and 3 are from cytoplasmic and vacuolar inorganic phosphate, respectively, and their chemical shifts can be used to monitor pH changes in each compartment of the cell. [Reproduced from Gerendas and Ratcliffe (2000) *J. Exp. Bot.* **512**: 207–219]

In this study live root tips were excised from 2 day-sprouted maize and placed in a 10 mm NMR tube with circulating oxygenated buffer. Spectra were run on a 300 MHz instrument. The roots were exposed to buffer containing ammonium sulfate at varying concentrations for 1 h, rinsed and the ^{31}P or ^{14}N-NMR taken as a time course. Figure 7.4 shows three such ^{31}P-NMR spectra of root tips taken before, at the end of 60 min of exposure to 10 mM ammonium sulfate, and after 3 h recovery. The spectra show definite changes; the spectra are very rich, well re-solved and many phosphorylated species are assignable to specific molecules. Some of the species present in the spectrum are glucose 6-phosphate, phosphocholine, γ-, α- and β-phosphates of nucleoside tripihosphates, uridene-diphosphate-glucose and NAD(P)H. As explained in the legend of Figure 7.4, both vacuole and cytosol pH can be determined from shifts of their respective inorganic phosphate lines.

Rather than follow the details of pH and ammonium metabolism of this experiment through, for the purposes of this chapter the reader needs only to understand that one can obtain a great deal of cellular detail from simple ^{31}P, ^{14}N, ^{15}N, or ^{13}C NMR spectra of whole or parts of living organisms.

7.3.3 Identification and quantification of small quantities in a complex mixture by 2D ^{1}H–^{31}P-HMQC-TOCSY NMR

The sea urchin system constitutes one of the best models for NE (nucelur envelope) assembly (Larijani *et al.*, 2000). Its primitive genome simplifies signalling pathways involved in the regulation of NE assembly. Following fertilization, the sea urchin sperm nucleus undergoes the breakdown of its inactive envelope. The disassembly of the lamina and the chromatin decondensation are followed by fusion of fertilized egg membrane vesicles (MVs). It is known that the cytoplasmic membrane vesicles (MV) participate in the formation of the NE but details of how and which molecules participate depend on identifying and quantifying the role of the various fractions and their components. The cytoplasmic membrane can be separated on a sucrose density gradient into fractions MV1, MV2, MV3 and MV4. The MVs are made up of phospholipids and it is a major analytical challenge to identify and quantify them.

Phospholipids are a major class of molecules in virtually all living organisims and play a role in most membranes. They consist of a 'head group', which has a phosphate group esterified to a choline or ethanolamine or any of a variety of other short molecules. The phosphate is also esterified to glycerol, which in turn is esterified to a 'tail' of two fatty acid chains, of which there are many types. The nature of the head group and fatty acid chains vary greatly over species and specific membranes. The phosphorus nucleus couples magnetically to any protons which are two to four bonds away with a spin–spin coupling constant of about 7 Hz.

A direct ^{1}H NMR approach to quantify and identify the specific phospholipids would be the first choice given the high sensitivity of ^{1}H-NMR, and indeed this method can and has been used. However, once the mixture of phospholipids becomes complex there can be considerable overlap of peaks from the various phospholipids as well as any other protonated components in an extract so that the spectrum rapidly becomes uninterpretably messy. (This can be partially alleviated by going to higher magnetic field strengths.) Direct ^{31}P-NMR has the advantage that, with decoupling of any protons, one expects one ^{31}P peak per specific phospholipid. However, there are several drawbacks: (1) samples of the minor phospholipids can be very small and, even with the high sensitivity of ^{31}P NMR, days of acquisition would be required to get even minimal quality ^{31}P signal; and (2) all of the ^{31}P signals from phospholipids come within a narrow 1 ppm range and the ^{31}P peaks can move with solvent conditions, generating uncertainty as to which ^{31}P signal is to be assigned to which phospholipid.

The above problems are solved by doing a specific 2D NMR experiment; an indirect heteronuclear correlation pulse program (HMQC), which generates a 2D spectrum with its *x*-axis as a proton NMR scale and a *y*-axis as a ^{31}P NMR scale: a peak appears at a (^{1}H,^{31}P) coordinate that identifies the bonded pair of atoms. The optimization of this experiment depends on timings which are governed by the

^{31}P–^{1}H coupling constant (J_{PH}), which are generally known quantities, 7 Hz in this case. The pulse program selects for observation only those magnetic components which involve *both* ^{1}H and ^{31}P quanta (multiple quantum, MQ) and correlates the specific ^{1}H signal with the ^{31}P to which it is coupled. This is the HMQC part of the pulse program. An additional section of the pulse program spreads the magnetization out among each separate chain of coupled protons. This means all the coupled protons of, for instance, a phosphotidylcholine head group would appear on a line with the ^{31}P signal of that group. This part of the pulse program is called TOCSY (total correlation spectroscopy). A projection of the results as a 1D plot along the *x*-axis gives the sum of all ^{31}P-coupled proton components in the sample. The projections looks exactly like a 1D proton spectrum; similarly, for the *y*-axis as a ^{31}P spectrum. However, each phospholipid is resolved along the ^{31}P axis: a trace parallel to the *x*-axis from its ^{31}P chemical shift position would show the resolved ^{1}H spectrum of that specific phospholipid. Thus a complex mixture is resolved and components identified via NMR without recourse to actual chemical separation. Each 'peak' in the 2D spectrum shows a ^{1}H–^{31}P correlation of a specific intensity at that cross peak. With proper calibration by seeding the sample with known quantities of a given phospholipid, each component can be quantified. The 'indirect' aspect of the experiment means that we do not directly observe the ^{31}P spectrum, but measure it indirectly through the protons that are coupled to the ^{31}P. This has the advantage that we observe the ^{31}P with the sensitivity of ^{1}H which is $2.3\times$ greater than direct ^{31}P observation. With the HMQC pulse program, the overall sensitivity is higher by a factor of $2.3^{2.5} = 8.2$ than for direct ^{31}P observation. Thus the increase per scan: to get to a given signal-to-noise with the indirect vs direct experiment would save a factor of $8.2^2 = 67$ in time, thus greatly increasing the sensitivity (or shortening the time) of the overall observation requirements. With standard diluted mixtures of phospholipids, reliable quantitative measurements could be made down to 0.04 μM in overnight experiments.

Figure 7.5 shows the HMQC-TOCSY 2D NMR spectrum of MV2 cleanly resolving the presence of L-α-phosphtidylcholine dipalmitoyl (DPPC), phosphotidylinositol dipalmitoyl (DPPI), DL-α-diphoshophotidyl-L-serine (DPPS) and DL-α-phosphotidyl-ethanolamine dipalmitoyl (DPPE). Each set of proton peaks for a given ^{31}P chemical shift constitutes a 'fingerprint' of the respective phospholipid. The volume of any of the 2D peaks (usually the stronger peaks are selected) can be measured and taken as the relative quantity of the given phospholipid in the mixture, providing one does a calibration of relative sensitivity. A very surprising result in this quantitative work was that another fraction, MV1, proved to have PI as its major (80 per cent) component. PI is a minor component in other fractions and so it is suspected that MV1 plays a special role in the NE formation. The role of PI is under further investigation.

There are various two-dimensional techniques for separating out specific interactions, connections or components besides HMQC. HMQC and HMQC-TOCSY are not restricted to ^{31}P–^{1}H interactions: more commonly they are used in ^{13}C–^{1}H cases, but in general, as is clearly illustrated in this example, 2D NMR spreads out resonances compared with 1D NMR so that desired components can be observed and quantified separately.

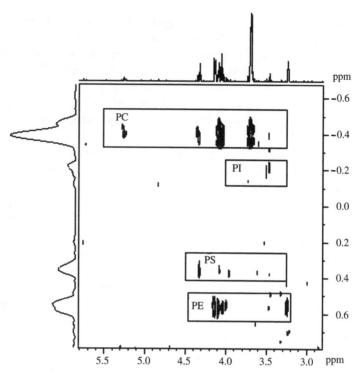

Figure 7.5 2D {1H-31P}-HMQC-TOCSY of membrane vesicle fraction of nuclear envelope of Lytechinus pictus as explained in text. The projection of the 2D peaks onto the left side of the figure gives a ^{31}P NMR spectrum. Along the top, the projection gives the ^{1}H NMR spectrum. [Reproduced from Larijani *et al.* (2000) *Lipids* **35**(11): 1289–1297]

7.3.4 Simultaneous separation and identification of very complex mixtures: LC–NMR

A recent exciting development of vast analytical importance has been LC-NMR – the combining of liquid chromatography techniques with NMR (Spraul *et al.*, 1993). In simplest terms, this means that one can separate mixtures by HPLC (high-pressure liquid chromatography) or other chromatographic techniques and run the output directly into a special capillary-flow NMR probe. LC-NMR probes are generally small diameter solenoid coils (some with a diameter as low as 200 μm) with a fine thin-walled tubing running through it so that LC effluent runs directly and continuously through the NMR as a detector. (The smaller the diameter of the NMR probe coil, the greater the sensitivity *per microgram* of sample.) Sample volumes can be as low as 10 μl for proton detection. Shown in Figure 7.6 is an example of use of NMR as a detector for HPLC; flow can be stopped for complete analysis of molecular structure if necessary by 2D NMR techniques. Thus, while UV as a detector only indicates an intensity, NMR can quantify and identify the molecular species eluting in a given fraction.

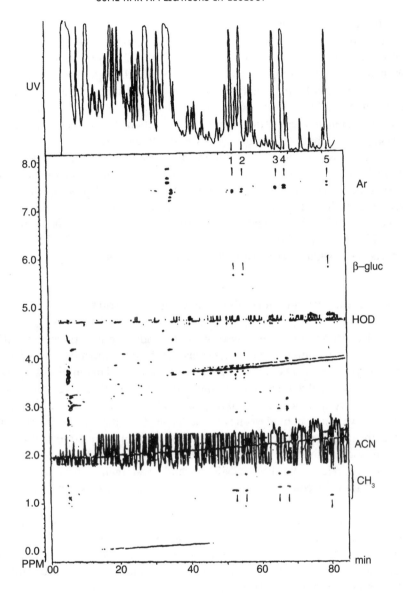

Figure 7.6 LC UV/NMR plot: the top part of the figure is the elution pattern of an HPLC column with UV intensity as the y-axis and elution time as the x-axis. The lower half of the figure plots NMR range vs elution time and the NMR intensity of the eluate is plotted as intensity coming out from the plane of the page. The intense peaks labeled HOD and A CN are respectively from the H_2O/D_2O and acrylonitrile elution solvents. The region labelled β-gluc shows specific ibuprofen metabolites. A vertical line through the bottom part of the figure would generate the NMR spectrum of the component eluting in that fraction with the NMR intensity coming out of the page. [Reproduced from Spraul *et al.* (1993) *Anal. Chem.* **65**: 327–330]

LC-NMR has allowed analysis and identification of components of mixtures to reach a level of unprecedented detail compared with prior methods. For example, some researchers have positively identified over 300 different molecular components of beer, including various oligosaccharides. Much effort has been devoted to biofluids and LC-NMR is contributing greatly to the new field of metabolomics, which is the multicomponent analysis of the effects of a stress/drug/toxin on a living system's metabolism. Metabolomics is the study of the effects of a stressor on an individual living system's metabolism; metabonomics is the larger multilcomponent statistical analysis of many systems to discern patterns of correlations of effects. The LC-NMR flow-through method allows for a more rapid analysis of a vast bank of samples. The current state of the art has 900 MHz proton spectra of individual samples (e.g. urine samples of a series of individuals) measured about every 20 s with a sample requirement of only 30 µl. Hundreds of molecules can be quantified for a given analysis. This means that thousands of individuals can be surveyed rapidly and easily and correlations sought among the vast amount of data generated.

7.3.5 Three-dimensional molecular structures in solution

By far the deepest and most impressively detailed advance of biology by NMR has been its use to determine three-dimensional molecular structures *in solution* of proteins, polynucleic acids and even polysaccharides. This molecular level of understanding has helped explain *at the molecular level* how certain enzymes promote their very specific bond breakings or makings, how cancer agents intercalate with DNA, how inhibitors bind to enzymes and a host of other biological phenomena.

A tractably simple example of the methods involved in NMR structural analysis is the determination of the structure of the peptide polyphemusin I, (Power *et al.*, 2004) sketched below. However, these NMR methods and their extensions have been applied to determine complete 3D solution structures of many hundreds of much larger molecules – proteins and many variants of DNA segments and of RNAs. The reader should consult the Further Reading list to do justice to this immense and elegantly detailed field. Often differences are found between NMR-determined structures and X-ray diffraction structures because NMR has a better handle on the flexible, not rigidly crystalline, parts of the molecule. Proteins with molecular weights less than or equal to 20 kDa can be fairly readily solved by NMR because the molecules tumble rapidly enough to give narrow NMR lines and there are not too many lines. More recent NMR techniques extend the molecular weight limit to about 100 kDa, but such large molecules require strenuous synthetic efforts in isotopic substitution; however, this area is advancing rapidly.

Polyphemusin I, a defensive agent of the Atlantic horseshoe crab (*Limulus polyphemus*), is an antimicrobial peptide composed of 18 amino acids with two cystine crosslinks. The amino sequence is readily determined to be RRWCFRVCYRGFCYRKCR-NH$_2$ by standard degradation techniques, but in order to understand the mechanism of how polyphemusin I kills bacteria, one needs to know the specific details of how it interacts with the membrane and other molecular parts of the cell. This molecule is specifically interesting because it appears to function not only by the usual mechanism of antibiotics by breaking cell walls, but it may also interfere with protein synthesis in the cell. Ultimately, understanding how polyphemusin I works will require the full 3D structure of the peptide.

The determination of a peptide 3D NMR structure is fairly straightforward, although there is a lot of necessary book-keeping of detail. In this case there is a great beauty in the discreteness of each resonance. The sample consists of a few milligrams of polyphemusin I in 0.5 ml of H_2O (+5 % D_2O for a lock signal). Several 2D proton NMR spectra are acquired using special techniques to suppress the water signal. The first spectrum is a 2D TOCSY. Any subset of protons that are sequentially coupled together, for example all the protons of a side chain of a given amino acid residue, show up on the same line. Part of the TOCSY spectrum is shown in Figure 7.6(a). The part of the x-axis shown is the proton NMR region of the peptide bond NHs. Along the y-axis are the chemical shifts of protons in the 0.5–5.5 ppm region. Which amino acid belongs to which NH chemical shift can be assigned by various methods but TOCSY itself identifies patterns of lines which are very characteristic to each amino acid. For example, the left-most set of peaks labeled R10 connected by the vertical line is characteristic of an arginine. There are 17 NH peaks in the 8.0–9.4 ppm region (one amino acid, R1, is the NH_2 end) and TOCSY helps identify which proton peaks belong to which amino acid, even resolving, for example, the six different arginines as well as the four cystines in polylphemusin I.

The second proton 2D spectrum used to discern 3D structure is NOESY, and is shown in Figure 7.5. The nuclear Overhauser effect (NOE) is a *through-space* interaction, which causes a proton to affect the intensity of any protons in its spatial vicinity. The NOE has a range of up to 0.6 nm. The effect can be quantified and in favorable cases actual distances between protons can be calculated. In the 2D form of the experiment, NOESY cross peaks show up between proton resonances that are near each other in space. This allows, for example, working through the amino acid sequence: the NH of Y14 interacts via the NOE to CH of C13, the NH of C13 with the CH of F12, and so on as can be traced through in the TOCSY–NOESY data. Also seen in the NOESY plot are a number of other interactions which are not just along the NH:CH sequence. Some of these are within the side chains and some are inter-residue. These distances constitute dimensional or proximity constraints on the 3D structure. In polyphemusin I, NMR yields 143 distance constraints: 69 intra-residue and 74 inter-residue.

With this constraint data, one then resorts to computer minimization of allowed conformations of the 18 amino acid chain and the results are show in Figure 7.7. Clearly much flexibility exists in certain bonds of the molecule, but other parts are very well defined by an anti-parallel β-sheet with a hairpin turn. There are 17 structures that fit the constraints and all are shown in the figure.

Importantly this structure reveals a cationic cleft in the molecule which would greatly aid in attaching this antimicrobial agent to phosphate moieties on membranes. While the determination of the mechanism is not complete, this study makes it clear how important it is to have full structures of molecules.

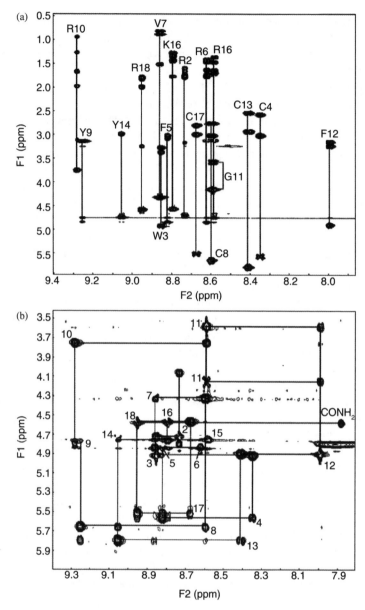

Figure 7.7   (a) NH region (*x*-axis) of the TOCSY ^{1}H-NMR spectrum of polyphemusin I with vertical lines added to emphasize connectivity to the α-CH and side chain region within each amino acid. Labels are single letter amino-acid abbreviations with the number indicating the position along the peptide chain, as given in the text. (b) NH-α-CH region NOESY spectrum of polyphemusin I showing interconnectivity of main chain sequence and some inter-chain interactions as detailed in the text, which delineate the three-dimensional structure of the peptide. [Reproduced from Powers *et al.* (2004) *Biochim. Biophys. Acta Proteins Proteomics* **1698**(2): 239–250]

(a)   (b)

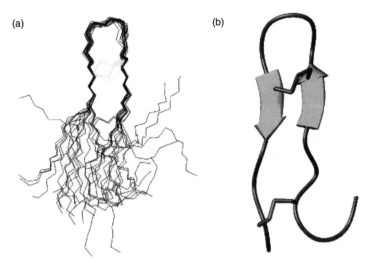

Figure 7.8 (a) Chain conformations of polyphemusin I consistent with the NMR determined constraints. The upper β-hairpin turn part is relatively rigid whereas the lower part has considerable conformational freedom. (b) Ribbon figure showing the β-sheet region of the peptide conformation. [Reproduced from Powers *et al.* (2004) *Biochim. Biophys. Acta Proteins Proteomics* **1698**(2): 239–250]

7.4 Conclusion

From these few examples it is clear that NMR has a very wide applicability in terms of nucleus observed or molecule or system studied. The reader is urged to expand this horizon with the material in the suggested Further Reading and to imagine variations on any one of the above examples for new applications.

Further reading

Gerendas, J. and Ratcliffe R.G. (2000). Intracellular pH regulation in maize root tips exposed to ammonium at high external pH. *J. Exp. Bot.* **512**: 207–219.

Larijani, B., Poccia, D.L. and Dickinson, L.C. (2000). Phospholipid identification and quantification of membrane vesicle subfractions by ^{31}P and ^{1}H two-dimensional nuclear magnetic resonance. *Lipids* **35**(11): 1289–1297.

Powers, J-P.S., Rozek, A. and Hancock, I.R.E.W. (2004). Structure–activity relationships for the β-hairpin cationic antimicrobial peptide polyphemusin I. *Biochim. Biophys. Acta Proteins Proteomics* **1698**(2): 239–250.

Rager, M.N., Binet, M.R., Ionescu, G. and Bouvet, O.M.M. (2000). ^{31}P and ^{13}C NMR studies of mannose metabolism in *Plesiomonas shigelloides*. *Eur. J. Biochem* **267**: 5136–5141.

Spraul, M., Hofmann, M., Dvortska, P. Nicholson, J.K. and Wilson, I.D. (1993). High-performance liquid chromatography coupled to high-field proton nuclear magnetic resonance spectroscopy: application to the urinary metabolites of ibuprofen. *Anal. Chem.* **65**: 327–330.

Fundamental NMR concepts and technique

Berger, S. and Braun, S. (2005). *250 and More NMR Experiments*. Wiley-VCH: Weinheim.

Claridge, T.D.W. (1999). *High-resolution NMR Techniques in Organic Chemistry*. Elsevier Science: Oxford.

Evans, J.N.S. (1995). *Biomolecular NMR Spectroscopy*. Oxford University Press: Oxford.

Reviews and edited symposia books on NMR applications

Beckmann, N. (ed.) (1995), *Carbon-13 NMR Spectroscopy of Biological Systems*. Academic Press: San Diego, CA.

Cavanagh, J., Fairbrother, W.J., Palmer III, A.G. and Skelton, N.J. (1996). *Protein NMR Spectroscopy* Academic Press: San Diego, CA.

Lindon, J.C., Holmes, E. and Nicholson, J.K. (2003). So what's the deal about metabonomics? *Anal. Chem.*, **75**(17): 384A–391A.

Markley, J.L. and Opella, S.J. (eds) (1997). *Biological NMR Spectroscopy*. Oxford University Press: Oxford.

Ratcliffe, R.G., Roscher, A. and Shachar-Hill, Y. (2001). *Plant NMR Spectroscopy. Prog. Nucl. Magn. Reson. Spectrosc.*, **39**: 267–300.

Smith, I.C.P. and Stewart, L.C. (2002). Magnetic resonance spectroscopy in medicine: clinical impact. *Prog. Nucl. Magn. Reson. Spectrosc.*, **40**: 1–34.

8

Solid-state NMR in biomembranes

Erick J. Dufourc

UMR 5144 CNRS, University Bordeaux 1, Institut Européen de Chimie et Biologie, Pessac, France

8.1 Introduction

The membrane that defines the whole cell or the cell organelles is in essence a medium that is half-way between liquid and solid. This state, called soft matter, is by definition a liquid crystalline medium whose anisotropic properties are essential for membrane function and further for cell life. Molecules embedded there, such as lipids or membrane proteins, may undergo many dynamic processes, such as lateral diffusion in the bilayer plane, transverse diffusion from one membrane leaflet to the other or conformational changes in the membrane interior. They may also group as patches in the membrane plane, named 'rafts' to illustrate the rigidity of these membrane domains that swim in a 'sea' of more fluid lipids and proteins. Under-standing the structure and dynamics of membrane components will not reveal the function of lipids or proteins, but will help in deciphering complex biological processes such as cell recognition, fusion, trafficking, apoptosis, energetics and signal transduction.

The purpose of this chapter is to show how nuclear magnetic resonance of the solid state is particularly suited to probing membrane structure and dynamics. NMR reports on magnetic properties of nuclei in atoms that are chosen because of a peculiar physical property called nuclear spin. Almost all nuclei of atoms commonly encountered in biological systems (H, N, C, O, P, F, Na, K, Ca, Mg, Fe) have a nuclear spin. The abundance of such isotopes may, however, vary from ca. 100 per cent

Chemical Biology Edited by Banafshé Larijani, Colin. A. Rosser and Rudiger Woscholski
© 2006 John Wiley & Sons, Ltd

(e.g. ^{1}H, ^{14}N, ^{31}P, ^{19}F, ^{23}Na, ^{39}K) to only a few per cent (e.g. ^{15}N, ^{13}C, ^{43}Ca, ^{17}O, ^{2}H). Some isotopes have no nuclear spin (e.g. ^{12}C, ^{16}O, ^{32}S) and cannot be used for NMR studies. NMR of the liquid state is a well-known powerful technique to determine the structure of molecules at atomic resolution in solution. Three-dimensional (3D) structures can also be obtained in the crystal state with X-ray or neutron diffraction. Unfortunately, the intrinsic physical state of the membrane where position order is lost but orientation order still there is not well suited either for high-resolution X-ray diffraction nor for NMR of liquids.

This chapter will develop the basic theory needed to understand how solid-state NMR can be utilized in membrane systems. In particular, it will be shown that wide-line NMR can be sensitive to molecular motions and is an appropriate tool to track membrane dynamics (membrane fluidity, fusion) and probe average orientations of molecules embedded in membranes (membrane topology). The 3D structure of molecules in membranes is also obtained by making use of magic angle sample spinning, a technique that leads to pseudo 'high-resolution' spectra as in the liquid state, the sample being still in the membrane 'liquid crystalline' state. Another facet of NMR is also presented, i.e. relaxation studies that allow measurements of the speed of molecular motion and activation energies, and allow membrane dynamics to be pictured from the atomic level where intra-molecular motions dominate to the cell level where membrane hydrodynamic modes of motion play an important role.

8.2   NMR basics for membrane systems

The following is restricted to membrane phases. However, the description is general enough that it stands for soft matter systems that possess internal disorder (liquid crystals, colloids, polymers, plastic crystals, etc.)

8.2.1   Anisotropy of NMR interactions in membranes

One-dimensional (1D) NMR spectra in liquids are made from very sharp lines with line widths less than 1 Hz. This is because the internal magnetic interactions that modulate the applied external magnetic field are averaged to scalars by fast (frequency 10^{12}–10^{15} Hz) isotropic tumbling of molecules in solution that averages the chemical shielding and the indirect spin–spin interactions. In liquid crystalline media, molecular motions are much slower and are anisotropic. The reorientation of molecules no longer occurs in all directions but happens, for instance, around the long axis of a lipid (frequency of 10^{9} Hz) or a peptide (10^{6}–10^{8} Hz) embedded in a membrane. This has two consequences: NMR spectra may span several thousand Hz and depend upon the orientation of molecules with respect to the magnetic field:

$$\omega = \omega_{\text{iso}} + A\left(\frac{3\cos^2\theta - 1}{2}\right)S \tag{8.1}$$

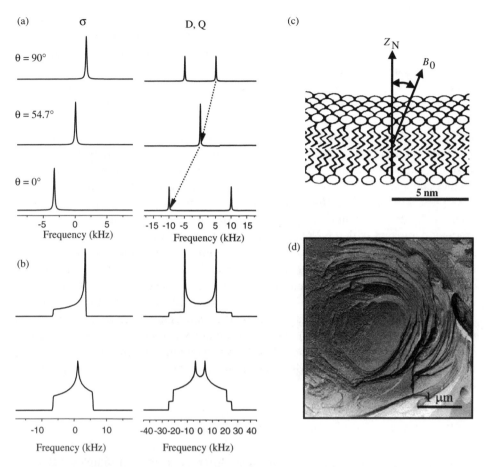

Figure 8.1 Solid-state NMR spectra of membrane media. (a) Spectra of samples being macroscopically oriented at the angle θ with respect to the magnetic field; σ (chemical shielding), D (direct spin-spin) and Q (electric quadrupolar) respectively stand for magnetic interactions that dominate spectra. (c) Schematic drawing of a membrane fragment whose bilayer normal (axis of motional averaging, Z_N) is oriented at an angle θ with respect to the magnetic field direction B_0. (b) Spectra of nonoriented samples ('powder patterns') for σ and Q interactions. The bottom spectra are for purely solid samples with no axial symmetry and the top spectra are for liposomes with fast axial rotation of molecules around Z_N. Powder spectra for D interaction are always axially symmetric, to first order. (d) Electron microscopy picture of freeze-fractured liposomes (multilamellar vesicles) leading to spectra of (c)

where ω is the frequency of a NMR line, θ the angle between the axis of fast motional averaging [lipid or peptide rotation axis for instance, Z_N in Figure 8.1(c)] and the magnetic field direction, B_0, and S the order parameter linked to membrane dynamics, which will be discussed later. ω_{iso} stands for the isotropic frequency as detected in purely isotropic liquids. The constant A depends upon the magnetic interaction of interest. Three internal magnetic interactions dominate spectra in membranes: the direct spin–spin or dipolar interaction (D), the chemical shielding interaction (σ) and the electric quadrupolar interaction (Q). The D interaction is found when two nuclei are close in space (a few Angstrom) and dominates ^{1}H spectra; σ mainly accounts for ^{31}P, ^{19}F and ^{15}N spectra, Q is only found for nuclei with a nuclear spin greater than $\frac{1}{2}$ that possess an electric quadrupolar moment (e.g. ^{2}H, ^{14}N, ^{17}O, ^{23}Na). ^{13}C spectra reflect the presence of both σ and D (^{1}H–^{13}C). The indirect spin–spin interaction (J) is so weak compared with the other interactions that it is barely detected in membrane spectra. As seen in Figure 8.1(a), the NMR lines shift upon change in orientation of the membrane plane (angle θ) with respect to the magnetic field. A doublet appears for D and Q interactions Figure 8.1(a); it is considered here that there is only one pair of spin $\frac{1}{2}$ nuclei interacting (^{1}H–^{1}H, ^{1}H–^{13}C) or one spin 1 nucleus (^{2}H, ^{14}N). This doublet collapses when the axis of motional averaging is oriented at the magic angle $\theta_m = 54.7°$ with respect to B_0. Liposomes [multilamellar vesicles, Figure 8.1(d)] or cells usually do not orient in magnetic fields. Because these large (μm) entities slowly reorient, compared with the NMR frequencies, all membrane orientations (θ varying from 0 to 90°) lead to a so-called 'powder spectrum'. The top traces of Figure. 8.1(b) are powder spectra for σ, D and Q interactions under conditions of axial symmetry as described by Equation (8.1), whereas the bottom traces stand for σ and Q interactions where there is no axial symmetry, as in the pure solid state. The theory describing such spectra is more complex than Equation (8.1) and is not given here for simplicity (see Mehring, 1976; Fyfe, 1983). As seen in Figure 8.1, solid-state NMR line shapes span tens-to-hundreds of thousands of Hz: this is 'wide line' NMR. They are also very different depending upon the NMR interaction that dominates and depending whether the sample is oriented or not.

8.2.2   Spectra and the effect of motional averages

As seen in Figure 8.1, there are considerable differences between samples that are in the solid state and those for which molecules undergo a fast axial rotation around the bilayer normal. Figure 8.2 ranks motions that may occur in membranes. It is seen that

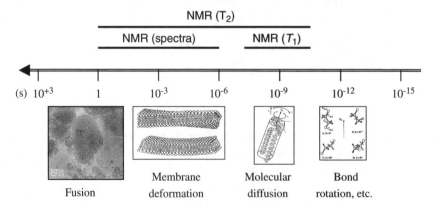

Figure 8.2   Motional time scale in biomembranes and NMR windows. Drawings and microscopy image depict the spatial scale at which events may occur

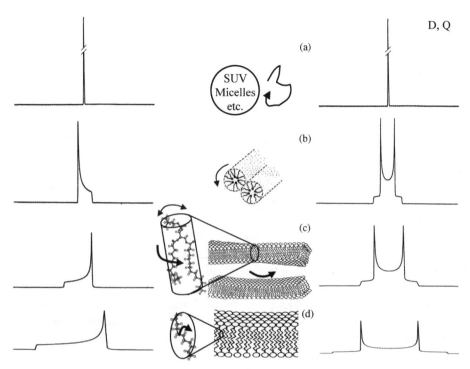

Figure 8.3 Motion-averaged solid-state NMR spectra for σ, D and Q interactions. Drawings picture the nature of motion being active: (d) fast axially symmetric intramolecular motions, (c) molecular and collective motions, (b) rotation of molecules around the cylinder axis of a hexagonal phase, and (a) isotropic motions

τ_c, the correlation time, which defines the time it takes for a molecule to reorient in a given membrane phase, can span ca. 20 time decades going from femtoseconds to hours. Motions that may be detected by NMR range from the picosecond to the second and can be intramolecular involving bond rotations, molecular such as rotation-diffusion of the entire molecule or collective, i.e. due to concerted diffusion of molecules in a bilayer (hydrodynamics). Both spectra and relaxation times (T_1, T_2) may bring dynamic information. One of the powers of wide-line solid-state NMR resides in the fact that motional processes with τ_c shorter than the reciprocal of the spectrum width will modify the observed line shape. Figure 8.3 pictures the effect of consecutive fast axial motions on NMR powder spectra; it has to be read from bottom to top, considering a hierarchy in time of motions, i.e. going from the fastest to the slowest. Intra-molecular motions [Figure 8.3(a)] typically occur in picoseconds (10^{-12}s) and average the static magnetic interactions (σ, D, Q) leading to axially symmetric lineshapes of smaller width. The molecular motions (nanoseconds, 10^{-9}s) produce further averaging that reduces further the width of lines [Figure 8.3(b)]. Collective motions that lead to membrane deformation over microns occur from nanoseconds to seconds. Some of these hydrodynamic modes of motion further reduce the line width. As a general statement, all anisotropic motions of τ_c shorter than the μs will reduce spectra width. This is simply translated in terms of less and less ordering, i.e. a reduction of the order parameter S of Equation (8.1). A special case is also shown in Figure 8.3(c), a fast motion that is present in a hexagonal phase that may be encountered in some fusion processes. Here the lipids and proteins are distributed along long tubes around which a fast rotation occurs. Because the new rotation axis makes an angle of 90° with respect to the former bilayer normal, this leads to a geometry-induced reduction by a factor of 2 of the spectrum width, compared with a bilayer-type spectrum. In the case of the σ interaction, this leads to an inversion in the symmetry of the line shape [compare Figure 8.3(b) and 3(c) left panel]. Finally, the effect of an isotropic motion is shown in Figure 8.3(d). The consequence of this is a complete cancelling of the angular dependent term of Equation (8.1), i.e. all solid-state interactions vanish. The spectrum is then that of a sample in the liquid state where only the isotropic σ and J interactions are present. It must be mentioned here that isotropic motions may occur in membrane phases, i.e. the local structure is organized (micelles, small vesicles, cubic phases, etc.), but the tumbling of small (1–30 nm) objects or the fast lateral diffusion on highly curved structures leads to spectra of liquid state NMR. Interestingly, for intermediate sizes (30–500 nm), the NMR line shape is in between that of liquids and solids and may be used for determining vesicle size.

8.2.3   A special case of motional averaging: magic angle sample spinning

Fast spinning around a molecular axis leads to efficient averaging of the solid state interactions. Andrew and Lowe placed the sample inside a tube that spins at several kHz around an axis oriented at $\theta = 54.7°$. This will in principle cancel the angular dependent term of Equation (8.1), the spectrum being left with the isotropic terms of σ and J interactions, as in pure liquids. How far lines can be narrowed depends on several factors. As a general statement, it can be said that one must spin the sample faster that the residual powder pattern produced by the residual solid–state interactions. This is where the limitation may be: D and Q interactions may lead to spectra that span over tens to hundreds of kHz. State-of-the-art technology affords magic angle spinning (MAS) speeds of a maximum of 50 kHz that may be in some cases insufficient.

8.2.4   Relaxations times: measuring dynamics

NMR possesses another facet, the nuclear relaxation times (T_1, T_2, etc.), which allows measurement of dynamic parameters such as correlation times, activation energies, diffusion constants, etc. Relaxation times are linked to τ_c through:

$$1/T_1^\lambda = K^\lambda[aJ(\omega_0) + bJ(2\omega_0)], \quad 1/T_2^\lambda = K^\lambda[cJ(0) + dJ(2\omega_0) + eJ(2\omega_0)],$$
$$J(p\omega_0) = \tau_c/(1 + p\omega_0^2\tau_c^2) \tag{8.2}$$

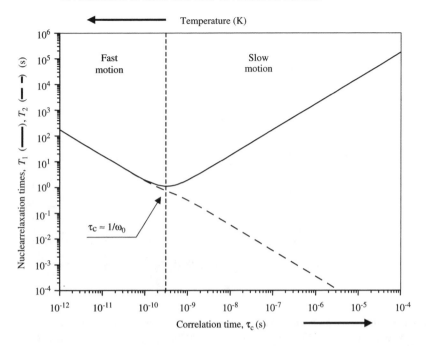

Figure 8.4 Relaxation times (T_1 and T_2) vs thermally activated motions. Correlation times and relaxation times are expressed in seconds on log scales. T_1 minimum delimitates the fast and slow motional regimes. Slopes directly lead to activation energies

where $\lambda = \sigma$, D and Q, K is a constant that contains time and space averages of the solid-state interactions, $\omega_0 = 2\pi\nu_0$ is the angular frequency associated with the static magnetic field in which experiments are run (typically, $\nu_0 = 500$ MHz) and a–e are numerical constants ($b = e = 0$ for $\lambda = \sigma$). Because several magnetic interactions may contribute to the observed relaxation, one may choose experiments in which σ, D or Q dominate the relaxation. Figure 8.4 shows typical data that can be obtained by recording T_1 and T_2 as a function of temperature. If a T_1 minimum is obtained in the range of temperatures explored, then the correlation time is obtained directly from the reciprocal of ω_0. On either side of $1/\omega_0$, i.e. in the fast and slow motion regimes, the K constants must be evaluated to obtain τ_c. Figure 8.4 shows that T_1 is a better reporter for nanosecond motions whereas T_2 is more sensitive to microsecond motions or slower. Because the motion is thermally activated, the activation energy, E_a, may be obtained using the Arrhenius equation ($\tau_c = A \exp\{E_a/k_B T\}$) and Equation (8.2).

8.3 Applications of wide-line NMR to membrane systems

The following examples show the use of wide-line NMR to probe structure and dynamics of lipids and peptides in membranes. Although not complete, the examples chosen represent the potential of the technique.

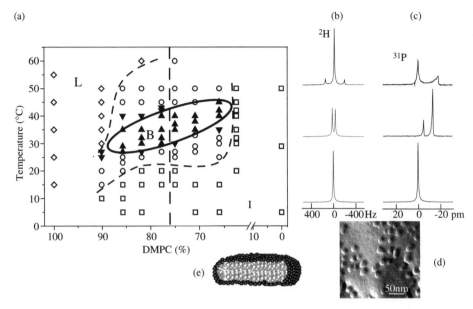

Figure 8.5   NMR and diagrams. (a) Temperature-composition diagram of DMPC(C_{14})–DCPC(C_6) systems at 80 per cent hydration (from Raffard *et al.* 2000). Symbols represent experimental points; letters stand for lamellar (L), bicelle (B) and isotropic (I) regions. (b) Selected ^{2}H-NMR spectra of heavy water along the dashed line on decreasing temperature. (c) As in (b) with ^{31}P-NMR. (d) Electron microscopy picture of the B region (Arnold *et al.* 2002). (e) Schematic drawing of a bicelle

8.3.1   Lipid phases: diagrams, peptide-induced fusion–fragmentation

As demonstrated in Figure 8.3, wide-line NMR powder spectra can effectively be used to determine which phase a lipid membrane might be in. This has been widely utilized to determine phase diagrams, e.g the phase diagram of dipalmitoylpho-sphatidylcholine (DPPC)–cholesterol in excess water (Vist and Davis, 1990) using deuterium NMR of chain-labeled DPPC, or that of phosphatidylethanolamine with and without cholesterol using ^{2}H and ^{31}P NMR. Figure 8.5(a) reports a temperature–composition diagram of short-chain and long-chain lipid mixtures (Raffard *et al.*, 2000). This system is interesting because it may form Lamellar phases (L), small discs (B) and isotropically tumbling micelles (I) depending upon composition and temperature. The small (30–50 nm) discs [Figure 8.5(e)] are called bicelles and orient naturally in the magnetic field [Figure 8.5(d)], leading to unambiguous spectral assignment using the naturally abundant phosphorus nucleus as a reporter [Figure 8.5(c)]. An isotropic line is obtained for isotropically tumbling micelles (0 ppm, bottom). Two sharp lines (−4, −10 ppm), corresponding to short-chain and long-chain lipids, are detected for discs normally oriented perpendicularly to the field (middle) and an isotropic line superimposed on a powder pattern, are seen for a biphasic region where lamellar and isotropic phases coexist (top). Areas of the two sharp lines can even be used to determine the bicelle size because they respectively

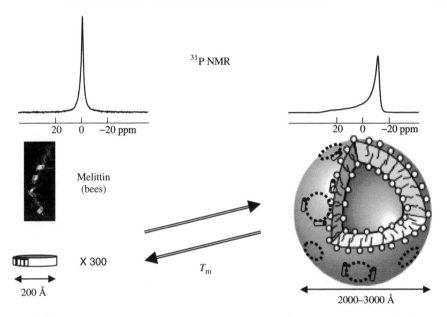

Figure 8.6 Melittin-induced membrane fusion–fragmentation process as followed by ^{31}P NMR. The amphipathic melittin 3D structure was obtained from the PDB using the ViewerLite software. Adapted from Dufourc *et al.* (1986) and Pott and Dufourc (1995). The vesicle drawing is courtesy of Dr Reiko Oda (IECB-CNRS)

correspond to lipids in the half tore and in the discoidal part. Deuterium NMR of deuterated water can also be used to obtain similar information [Figure 8.5(b)]. Note that in the bicelle domain the oriented spectrum is composed of a unique doublet. This indicates that, on average, all the water in the sample is ordered. This property is used in solution NMR to measure residual dipolar couplings that allow the determination of the 3D structure of soluble proteins dissolved in this ordered water system . Phosphorus powder spectra can also be used to monitor the effect of toxic amphipathic peptides such as melittin (from bee venom) or haemolysin (secreted by *S. aureus*) on biomembranes . Toxin-induced membrane fragmentation–fusion can be seen in Figure 8.6, following the isotropic line (small, 200 Å, tumbling objects) that melts, upon heating into a nonregular powder pattern, reflecting the formation of large (2000–3000 Å) vesicles being deformed by the magnetic field.

8.3.2 Bilayer internal dynamics: order parameters, membrane thickness, sterols

Besides monitoring morphological changes on membranes, wide-line NMR can also report on internal membrane dynamics, i.e. determine the microfluidity of the bilayer core. This is where the concept of order parameter takes place. Order parameters, S, are time and space averages (denoted by angular brackets) of the quantity

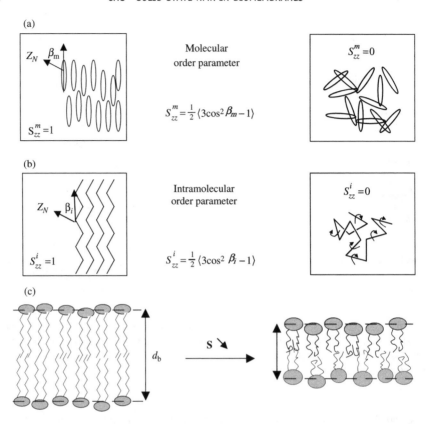

Figure 8.7 Order parameter concept. (a) Molecular order; (b) intramolecular (bonds) order: (c) correlation between order and membrane thickness, d_b: left, ordered membrane with little bond or molecule fluctuations (large d_b; right, less ordered membrane with bond and molecule fast reorientations within the membrane (small d_b)

$\langle 3\cos^2\beta{-}1\rangle/2$. They can be expressed at the intramolecular level, describing the fluctuation of bonds, at the molecular level, picturing the angular reorientation of molecules, or at the mesoscopic scale, reflecting membrane deformations. Bond rotations, *gauche-trans* isomerizations of lipid or peptide chains may be described as angular excursions of a local bond axis with respect to the bilayer normal. The quantity $S_{zz}^i = \langle 3\cos^2\beta_i - 1\rangle/2$ will then range from 1 for no bond fluctuation to 0 for liquid-like isotropic chain disorder [Figure 8.7(b)]. Similarly, lipid or peptides may reorient as a whole in the bilayer, their long axis orientation fluctuating with respect to the bilayer normal. $S_{zz}^m = \langle 3\cos^2\beta_m - 1\rangle/2$ also ranges from 1 to 0, as pictured in Figure 8.7(a). Order parameters are therefore a simple method to show if a system is rigid (S close to 1) or liquid-like (S close to 0). As indicated in Equation (8.1), S is directly measured from the spectrum width. It is in fact the product of intramolecular, molecular and collective order parameters. The best interaction to deal with when measuring order parameters is the quadrupolar interaction. Figure 8.8(a) shows a deuterium wide-line bicelle spectrum of ^{2}H-chain labelled DMPC

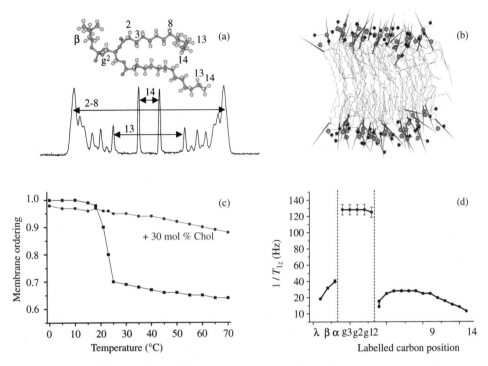

Figure 8.8 Measuring order parameters with ^{2}H-NMR. (a) ^{2}H-NMR wide-line spectrum of chain lablled DMPC embedded in a bicelle oriented parallel to the magnetic field and molecule display with carbon numbering. Measurement of quadrupolar doublets (arrows) indicates that the chain ends (C_{14}) have little order compared with positions near the backbone (C_{2-8}). (b) Snapshot of a molecular dynamics calculation showing the disordering of the interfacial head groups and of the bilayer centre. (c) Thermal variation of the DMPC molecular order parameter in the absence and presence of 30 mol per cent cholesterol. (d) Relaxation rate ($1/T_1$) as a function of carbon labelling (a) in perdeuterated DMPC embedded in a bicelle. This pictures the dynamics of a half-bilayer showing high mobility of both the chains and the head group and highly restricted dynamics of the lipid glycerol backbone. Adapted from Aussenac *et al.* (2003) and Douliez *et al.* (1996)

together with the molecule showing the positions of labelling. Several doublets are seen that are directly proportional to the order parameter S. Peak assignment was made based on selective labeling. From the spectrum it can be seen that the chain ends are very mobile (small doublet = small S) and that positions 2–8 display the same ordering. Relaxation times, T_1, can also report on membrane dynamics. Figure 8.8(d) shows how $1/T_1$ varies with lipid position in a membrane. It is seen that both the acyl chains and the head group are quite dynamic, particularly towards the chain ends. In contrast, the glycerol backbone that delimites the membrane interface has very low dynamics, acting as a knee–cap separating two fluid zones, as also seen in a molecular dynamics snapshot [Figure 8.8(b)]. Perturbations of membrane ordering can then be followed both at the membrane surface and down in the bilayer interior and allow monitoring membrane capping/insertion of peptides. Figure 8.8(c) shows how the molecular order parameter may be used to picture a gel-to-fluid transition in synthetic or natural (sphingomyelin for instance) membrane lipids. At low temperatures the molecules barely move in the membrane, $S_{zz}^m = 1$, whereas disorder occurs above 24°C (T_m of DMPC), S_{zz}^m, reaching 0.65. In the presence of 30 mol% cholesterol, the lipid molecular order increases considerably in the fluid phase, reaching values greater than 0.9, whereas at low temperatures, cholesterol decreases slightly the lipid ordering. The same ordering information can also be obtained from labelled cholesterol. ^{2}H-labelled molecules are nonperturbing probes and can be incorporated in natural membranes to report on their fluidity and detect or not the presence of lipid rafts . Because the membrane ordering is directly linked to bond or molecule space averages, it can be translated, using an appropriate theory, to average length of a molecule in a membrane and hence to membrane thickness [Figure 8.7(c)]. It is shown that increase/decrease in bilayer order means increase/decrease in membrane thickness: 30 mol% cholesterol leads to a 7 Å increase in the DMPC bilayer thickness.

8.3.3   Orientation of molecules (sterols, helical peptides) in membranes

Equation (8.1) indicates that frequency lines depend upon the orientation of molecules with respect to the magnetic field. This is used to determine the average orientation of molecules in the membrane if one knows the orientation of the membrane plane with respect to B_0 and atomic coordinates of the molecule of interest. Whereas bicelle membranes naturally orient in the field, classical membranes must be sandwiched in between glass plates that are macroscopically oriented by means of a goniometric device. The orientation of ^{2}H-labelled cholesterol in membranes has thus been determined by combining neutron coordinates and wide line deuterium NMR. Figure 8.9(a) shows the result: cholesterol is vertical in the membrane, offering the minimum molecular area to optimize lipid packing. ^{15}N-labelled peptides that are otherwise known to be helices have also been studied. Amide group ^{15}N–H bonds of the peptide backbone structure are approximately collinear with the helix axis and will give oriented spectra at 70 ppm if the helix lies flat in the bilayer or at 200 ppm if it is vertical, perpendicularly embedded in the membrane. Use of both ^{2}H and ^{15}N wide-line NMR on oriented samples has elucidated the topology of several helical peptides in membranes (gramicidins, Vpu protein, M2 peptide).

8.3.4   Membrane dynamics from picoseconds to milliseconds

From measurement of relaxation times, a detailed analysis of intramolecular, molecular and collective motions that occur in a membrane with temperature change may be observed, Figure 8.10. Correlation times are plotted in a log scale as a function of the reciprocal of temperature in order to have a direct reading of activation energies from the slopes of the lines; the steeper the slope, the higher the E_a. In the gel phase ($L_{\beta'}$, $P_{\beta'}$) there are only intramolecular ($\tau_{1,2,3}$) and molecular motions ($\tau//,\perp$) whose τ_c ranges from the ms to the ns. In the fluid, L_α, phase, above T_m, there

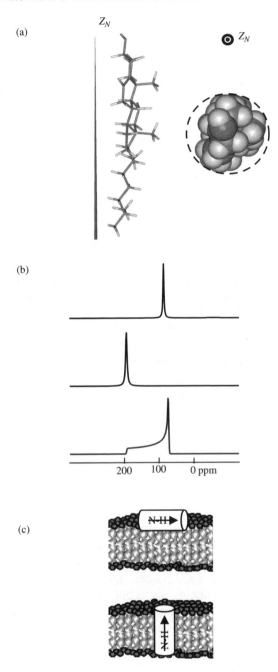

Figure 8.9 Orientation of molecules in membranes from NMR. (a) ^{2}H-cholesterol orientation with respect to the membrane normal, Z_N, seen from the plane (left) and from above (right), from Aussenac *et al.* (2003). (b) ^{15}N-NMR spectra of peptide-containing membranes oriented perpendicularly to the magnetic field. (c) Schematics of ^{15}N-labelled helices oriented flat on the membrane or embedded normal to the membrane plane, with corresponding spectra on (b), top and middle. The Bottom spectrum is that of nonoriented peptide

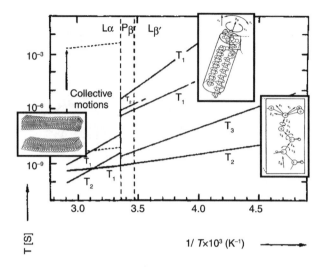

Figure 8.10   Correlation times vs reciprocal of temperature for DMPC membranes. Data is obtained from relaxation time measurements (Dufourc *et al.*, 1992). Drawings in inserts picture intramolecular ($\tau_{1,2,3}$), molecular ($\tau_{//,\perp}$) and collective motions, the latter only occurring in the fluid L_α phase. The Gel phase is represented by P_{β}' and L_{β}' regions

is an increase by three orders of magnitude of the motion speeds. Intramolecular motions go into the ps range that is barely detected by NMR. Collective modes of motions (membrane large-scale deformation) come into play only in the fluid phase. Addition of cholesterol (not shown) to this system levels off the large temperature-induced change in τ_c at T_m.

8.4   Applications of MAS to biomembranes and natural colloids

Liquid state NMR leads to atomic resolution structures provided there are sharp lines that (i) can be assigned to nuclei of atoms in molecules and (ii) can be related through space, showing the primary, secondary and tertiary structures of proteins. In the soft-matter state, there are no longer sharp lines, unless the static electric and magnetic interactions are removed by spinning the sample at the magic angle. This technique leads to spectra that are close in sharpness to those of the liquid state, and is termed high-resolution-MAS as opposed to purely solid-state cross polarization (CP) MAS techniques that operate on dry powders, polymers and amorphous crystals, and lead to less well-resolved spectra.

8.4.1   Three-dimensional structure of peptides in membranes

Proton NMR remains the best technique to obtain 3D structure of peptides and proteins. In a membranous environment, it may, however, be difficult to reach

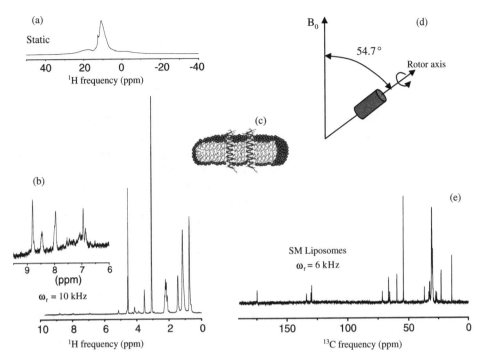

Figure 8.11   HR-MAS on membranes. (a) Nonspinning ^{1}H-NMR spectrum of a DMPC/DCPC bicelle sample containing the 35 amino acid transmembrane peptide of the tyrosine kinase receptor neu/ erbB-2. (b) The same as (a) except for MAS spinning at 10 kHz; the insert is an expansion of the NH region showing peptide resonances adapted from Sizun *et al.* (2004). (c) Schematic representation of the peptide embedded in a bicelle membrane. (c) Schematic representation of the MAS setup: the sample is inside a 4 mm rotor (12–100 µL) that spins at an angle of 54.7° (the magic angle) with respect to the magnetic field B_0. The carbon-13 spectrum HR-MAS of Sphingomyelin liposomes, MAS rate = of 6 kHz is courtesy of Cécile Loudet, CNRS-IECB

conditions of high resolution because of the strong dipolar homogeneous interaction (^{1}H–^{1}H) that may lead to spectra several tens of kHz wide. Figure 8.11(a) shows the nonrotating ^{1}H-NMR spectrum of a system composed of lipids, peptides and water (90 per cent) that spans ca. 60 ppm. Rotating the sample at the magic angle [Figure 8.11(d)] brings a pseudo high-resolution spectrum [Figure 8.11(b)] dominated by the lipid resonances between 0 and 5 ppm. The peptide NH resonances that are so useful to obtain the peptide 3D structure are nonetheless detected around 9–7 ppm (see insert expansion) with a lower intensity due to a relatively low amount of peptide inserted in the bilayer. Although the lipid resonances may not perturb the assignment experiments using multidimensional NMR techniques, they may be removed using perdeuterated synthetic lipids or natural lipids that have been grown on deuterium– enriched media. What resolution can one obtain under such conditions? It varies from 40 to 5 Hz depending on the type of membrane used, i.e. one or two orders of magnitude less than that of liquid-state NMR. It appears that resolution is linked to several factors, the most important being the membrane internal dynamics. Using bicelles [Figure 8.11(c)] that may possess low-order parameters, a 5 Hz resolution is possible for the lipid resonances. The linewidth of embedded peptide resonances depends on the intrinsic peptide dynamics in the membrane. If the peptide has little internal motion, MAS spinning at 10–30 kHz will not be sufficient to obtain the 3D structure of peptides in membranes using ^{1}H-NMR. Carbon-13 or nitrogen-15 NMR may then be used because the interactions averaged by MAS techniques are much less important than with ^{1}H-NMR. Spinning at moderate speeds (ca. 10 kHz) produces high–resolution spectra [Figure 8.11(e)] that are very close to those obtained in liquids. The advantage of ^{13}C-NMR is the spread of isotropic chemical shifts on 200 ppm. The disadvantage is the low natural abundance of ^{13}C (1.1 per cent) and ^{15}N (0.37 per cent) nuclei in samples, which can be circumvented by chemical or biological labelling of molecules. The 3D structure of a fully labelled protein in an extreme situation, i.e. in the form of dry powder, has recently been solved by magic angle spinning and multidimensional NMR techniques.

8.4.2   Three-dimensional structure of wine tannin–salivary protein colloidal complexes

The example of 3D structure determination in the aggregated form in Figure 8.12 shows the structure of a colloidal complex of wine tannins and human saliva protein. As in membranes, the colloidal form [large particles, Figure 8.12(a)] prevents high-resolution spectra. Proton HR-MAS, without labelling, together with the battery of multidimensional techniques of liquid-state NMR, permits the structure of the complex and its stoichiometry to be determined [Figure 8.12(b)]. Diffusion experiments performed with an HR-MAS probe equipped with z-gradients also enables the measurement of the hydrodynamic radius of the colloidal particles.

8.4.3   Distance determination using MAS recoupling techniques

The A factor of the D interaction [Equation (8.1)] is inversely proportional to the third power of the spin–spin distance, so measuring the doublet separation (Figure 8.1), leading to direct distance determination. Unfortunately, MAS cancels the solid-state interactions, and distance information between interacting nuclei can no longer be measured. There is, however, a possibility of retrieving the distance information if one spins the sample at a speed that is a multiple of the doublet distance.

8.5   Conclusion

Solid-state NMR used on biomembranes has many potential applications. Wide line NMR, MAS techniques and relaxation time measurements give information on

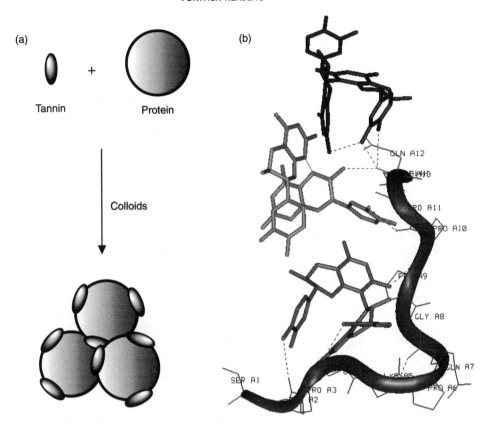

Figure 8.12 3D structure of colloidal complexes. (a) Schematic representation of the colloidal complex made from wine tannins and human saliva proteins. Complexation leads to aggregation that is otherwise known to be the origin of wine astringency. (b) Structure of the complex (IB7$_{14}$ protein and 3 B3 tannin molecules) as obtained from ^{1}H HR-MAS NMR, spinning at 5 kHz. The protein backbone is blue and the tannins green, pink and orange. Adapted from Simon *et al.* (2003)

membrane morphology, dynamics, topology and structure by taking advantage of the intrinsic anisotropy of the medium. The use of MAS reveals the 3D structure of peptides and proteins within membranes that cannot be otherwise obtained by liquid-state NMR or crystallography. Because this technique applied to biological complex systems is still in its infancy, one can foresee many more applications in the near future.

Further reading

Arnold, A., Labrot, T., Oda, R. and Dufourc E.J.: (2002). Cation modulation of 'bicelle' size and magnetic alignment as revealed by solid state NMR and electron microscopy. *Biophys. J.* **83**: 2667–2680.

Aussenac, F., Tavares, M. and Dufourc E.J. (2003). Cholesterol dynamics in membranes of raft composition: a molecular point of view from ^{2}H and ^{31}P solid state NMR. *Biochemistry* **42**: 1383–1390.

Creuzet, F., McDermott, A., Gerhard, R., Van Der Hoef, K., Spijkner-Assink, M.B., Herzfeld, J., Lugtenburg, J., Levitt, M.H. and Griffin R.G. (1991). Determination of membrane protein structure by rotational resonance NMR: bacteriorhodopsin. *Science* **251**: 783–786.

Douliez, J.P., Bellocq, A.M. and Dufourc, E.J. (1994). Effect of vesicle size, polydispersity and multilayering on solid state ^{13}P- and ^{2}H-NMR spectra. *J. Chim. Phys.* **91**: 874–880.

Douliez, J.P., Léonard, A., Dufourc E.J. (1995). A restatement of order parameters in biomembranes. Calculation of C–C bond order parameters from C–D quadrupolar splittings. *Biophys. J.* **68**: 1727–1739.

Douliez, J.P., Léonard, and A. Dufourc E.J. (1996). Conformationnal approach of DMPC sn-1 versus sn-2 chains and membrane thickness: an approach to molecular protrusion by solid state ^{2}H-NMR and neutron diffraction. *J. Phys. Chem.* **100**: 18450–18457.

Dufourc, E.J., Smith, I.C.P. and Dufourcq, J. (1986a). Molecular details of melittin-induced lysis of phospholipid membranes as revealed by deuterium and phosphorus NMR. *Biochemistry* **25**: 6448–6455.

Dufourc, E.J., Faucon, J.F., Fourche, G., Dufourcq, J., Gulik-Krywicki, T., Maire, M. Le. (1986). Reversible disc-to-vesicle transition of melittin-DPPC complexes triggered by the phospholipid acyl chain melting. *FEBS Lett.* **201**: 205–209.

Dufourc, E., Dufourcq, J., Birkbeck, T.H. and Freer J.H. (1990). δ-Haemolysin from staphylococcus aureus and model membranes. A solid state ^{2}H NMR and ^{31}P NMR study. *Eur. J. Biochem.* **187**: 581–587.

Dufourc, E.J., Mayer, C., Stohrer, J., Althoff, G. and Kothe G. (1992). Dynamics of phosphate head groups in biomembranes. A comprehensive analysis using phosphorus-31 nuclear magnetic resonance lineshape and relaxation time measurements. *Biophys. J.* **61**: 42–57.

Marassi, F.M., Ma, C., Gratkowski, H., Straus, S.K., Strebel, K., Oblatt-Montal, M., Montal, M. and Opella S.J. (1999). Correlation of the structural and functional domains in the membrane protein Vpu from HIV-1. *Proc. Nat. Acad. Sci. USA* **96**: 14336–14341.

Marinov, R. and Dufourc E.J. (1995). Cholesterol stabilizes the hexagonal type II phase of 1-palmitoyl-2-oleoyl *sn* glycero-3-phosphoethanolamine. A solid state ^{2}H and ^{31}P NMR study. *J. Chim. Phys.* **92**: 1727–1731.

Marinov, R. and Dufourc E.J. (1996). Thermotropism and hydration properties of POPE and POPE-cholesterol systems as revealed by solid state ^{2}H and ^{31}P-NMR. *Euro. Biophys. J.* **24**: 423–431.

Pott, T. and Dufourc E.J. (1995). Action of melittin on the DPPC-Cholesterol liquid-ordered phase: a solid state ^{2}H and ^{31}P-NMR study. *Biophys. J.* **68**: 965–977.

Raffard, G., Steinbruckner, S., Arnold, A., Davis, J.H. and Dufourc E.J. (2000). Temperature-composition diagram of dimyristoyl-dicaproyl phosphatidylcholine 'Bicelles' self-orienting in the magnetic field. A solid state ^{2}H and ^{31}P-NMR study. *Langmuir* **16**: 7655–7662.

Simon, C., Barathieu, K., Laguerre, M., Schmitter, J.M., Fouquet, E., Pianet, I. and Dufourc E.J. (2003). 3D structure and dynamics of wine tannin-saliva protein complexes. A multitechnique approach. *Biochemistry* **42**: 10385–10395.

Sizun, C., Aussenac, F., Grelard, A. and Dufourc E.J. (2004). NMR methods for studying the strcuture and dynamics of oncogenic and antihistaminic peptides in biomembranes. *Magn. Reson. Chem.* **42**: 180–186.

Smith, S.O. and Bormann B.J. (1995). Determination of helix-helix interactions in membranes by rotational resonance NMR. *Proc. Nat. Acad. Sci. USA* **92**: 488–491.

Vist, M.R. and Davis, J.H. (1990). Phase equilibria of cholesterol dipalmitoylphosphatidylcholine mixtures: ^{2}H-nuclear magnetic resonance and differential scanning calorimetry. *Biochemistry* **29**: 451–564.

9

Molecular dynamics

Michel Laguerre

Institute Européen de Chimie et Biologie, Pessac, France

9.1 Introduction

Molecular simulation techniques were pioneered in the early 1970s by Peter Kollman, Martin Karplus, William Jorgensen and Norman Allinger. Thirty years on, this technique has developed enough to be considered as a routine analytical tool alongside NMR or other spectroscopic methods. The technique is used to predict how molecular fragments might behave dynamically and, unlike most classical methods, dynamic behaviour can be analysed at an atomic level. One of the major advantages of this technique is that the predictions produced can be used to inform further experimental investigations.

Owing to the increasing availability and constantly improving processing power of computers, access to software performing molecular dynamics calculations has become possible for both small systems and an isolated protein. A small cluster of 16 or 32 PC processors using Linux has the processing power equivalent or superior to the largest supercomputers of 10 years ago at a cost similar to an NMR probe, meaning that this technique is becoming one of the cheapest of all analytical methods. Moreover free (or extremely cheap) open-source software is readily available for Linux platforms (or even for Windows machines) like, for instance, CHARMM, GROMACS, GROMOS, NAMD, DL-POLY and TINKER.

9.2 The basis of molecular mechanics

Molecular mechanics is based on a simplified structure of matter. Atoms are considered as spheres with the mass of the corresponding atom and bonds regarded as springs; classical mechanics can be applied to this system, hence the term 'molecular mechanics' (Figure 9.1).

Chemical Biology   Edited by Banafshé Larijani, Colin. A. Rosser and Rudiger Woscholski

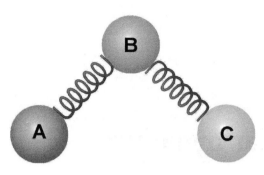

It is important to note that this model does not consider the presence of electrons or electron orbitals and for this reason, molecular mechanics does not allow access to reaction pathways involving bond making and breaking. The latter would require the use of quantum mechanics. On the other hand, if you are interested in structure, conformational space or dynamic behaviour of bio-molecules, then molecular mechanics is a reliable technique.

Figure 9.1   Schematic representation of the structure of matter within molecular mechanics

9.2.1   Force fields

In molecular mechanics the force parameters of the various springs found in organic systems have to be considered. All this data is stored by the software in a file termed the force field. Attempts to produce such data began in the early 1970s. For simple molecules these are called class I force fields and for more complex systems they are termed class II. A completely generalized force field for any molecule is not really available and it is necessary to choose a force field adapted to your system from those available.

Class I

- CHARMM

- AMBER

- MM2

- OPLS/AMBER

Class II

- CFF95

- MMFF94

- MM3 and MM4

- DREIDING

9.2.2 The energy problem

Next the behaviour of bonds within the system has to be addressed and used to determine a system's energy. Bonds can stretch, bend and twist. In addition there can be intermolecular forces such as Van der Waal's, electrostatic interactions and hydrogen bonding. In molecular mechanics the global energy of the system is the sum of six different terms: three 'bonded' terms and three 'nonbonded' (Box 1). The bonded terms comprise bond energy (interaction 1–2), valence angle energy (interaction 1–3) and torsional angle energy (interaction 1–4). Roughly, all these terms are proportional to the number of atoms in the studied system.

The nonbonded terms comprise the Van der Waal's (VdW) interactions (calculated from the Lennard–Jones potential), the electrostatic interaction (calculated from Coulomb's law) and eventually the H-bond interaction (calculated by a modified Lennard–Jones potential).

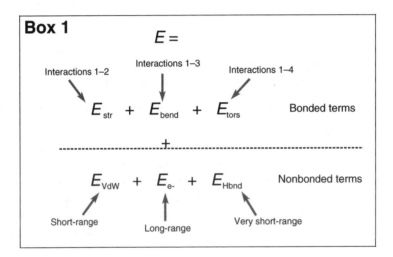

As an example, AMBER is a commonly used force field and with a relatively straightforward set of formulae (Box 2). In this equation the index '0' means reference values found in the force-field. For instance, for bond and valence angle interactions, the term is proportional to the square of the difference between the actual value and the ideal one. The torsional term is a little more complicated on account of the presence of a stable *trans* conformation and two metastable gauche conformations ($+60°$ and $-60°$). The VdW potential term arises directly from the Lennard–Jones equation with an attraction proportional to $1/r^6$ and a repulsion proportional to $1/r^{12}$. The electrostatic interaction, found from Coulomb's law, is the product of partial charges divided by the distance (note the presence of the dielectric constant ε which allows for the screening of electrostatic interactions by the polar solvent). Finally the H-bond terms are calculated from a modified Lennard–Jones potential with the $1/r^6$ term replaced by $1/r^{10}$, which implies interactions at a very short range: H-bonding is negligible beyond 4 Å.

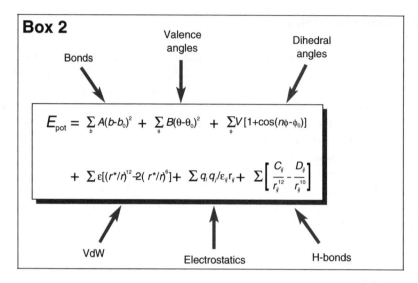

The balance between the two main terms of the nonbonded interactions (VdW and Coulomb) has a dramatic influence on the overall behaviour of any studied systems. This is a direct consequence of the nature of these interactions.

For organic chemists, even if the nonbonded terms are present, they have little influence on the overall structure of small organic compounds. In this field, the geometric parameters (bonded terms) are predominant.

In supramolecular chemistry and assemblies of small molecules such as solvents, liquid crystals, lipid mixtures and colloids, the structure and behaviour of the system is a direct consequence of the nonbonded terms. In these cases the important part of the VdW interaction is attraction (Figure 9.2).

In biological systems, like large proteins, the bonded terms are important locally but at longer distances the nonbonded ones become predominant. The balance between the geometric, VdW and Coulomb terms may be extremely complicated and the behaviour will be unpredictable.

Typically, Van der Waal's forces have a short range and weak interactions (see Figure 9.2). On the other hand, Coulomb interactions have a very long range and are very strong (see Figure 9.2).

If, in the studied system, there are only weak dipoles, then the electrostatic interactions will be almost negligible: the system is a *VdW phase*. In this case the system will be highly dynamic but extremely unorganized due to the weak and short-range interactions.

If, in the studied system, dipoles are strong or if charges are really present, then the electrostatic interactions will become predominant: the system is a *Coulombic phase*. In this case, the system will be poorly dynamic but highly organized.

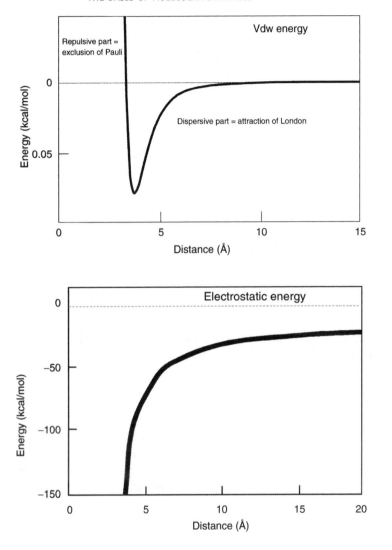

Figure 9.2 Analytical representation of the two main non-bonded interaction forms

The main problem in a biological system is that the nonbonded terms are roughly proportional to the square of the number of atoms. Thus, in the case of a small membrane protein embedded in a natural membrane with a typical number of atoms of around 25 000, one must compute roughly 625 000 000 Van der Waal's and Coulomb interactions.

9.3 The basis of molecular dynamics

Molecular dynamics builds on the ideas of molecular mechanics by simply adding random velocities (speeds) to the spheres (Figure 9.3). Newton's second equation

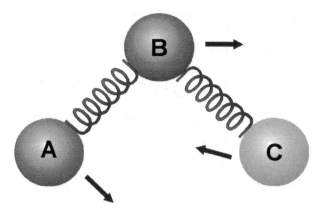

Figure 9.3   Schematic representation of the idea behind the molecular dynamics concept

of motion can now be applied to this system by integrating it with small time-steps, Δt:

$$F_i(t) = m_i \cdot a_i(t) \quad \text{(Newton's equation)}$$

Velocities are applied randomly on atoms (spheres) according to the temperature chosen for the experiment (see below). The overall technique is extremely simple but problems with two parameters have to be addressed. The first one is Δt: in order to integrate precisely a system exhibiting a large number of very different vibrations, it is necessary to integrate faster than the fastest of all vibrations. Biological systems are essentially organic systems so very light hydrogen atoms are present in large quantities implying very fast X–H vibration, on the order of 10^{-14} s. Thus to integrate accurately, Δt must be around 10^{-15} s (1 fs), i.e. representing a vibration that is 10 times faster than that of X–H. Hence a simple simulation lasting 1 ns requires 10^6 integrations of Newton's law. The second parameter is the energy term. In a small biological system with 25 000 atoms, a 1 ns dynamic run implies 10^6 integrations, each of them including the calculations of 625×10^6 nonbonded terms of *both* Van der Waal's and Coulomb effects (see above). To simplify the calculations it is common to neglect or approximate the nonbonded interactions for widely separated atomic pairs in order to lower the number of interactions included in the simulation. Nevertheless, this kind of approach is extremely CPU-demanding and generally necessitates the use of very powerful multiprocessor computers for long times. For example, on the most powerful processor available, a system containing 25 000 atoms needs around 800 h of computational time for a simulation of 1 ns.

9.3.1   Influence of temperature

Velocities are allocated randomly to atoms to produce a Maxwell–Boltzman distribution (Figure 9.4). If a low temperature is chosen (for instance 100 K, blue curve) then most atoms will have roughly the same velocity. In terms of the model at the

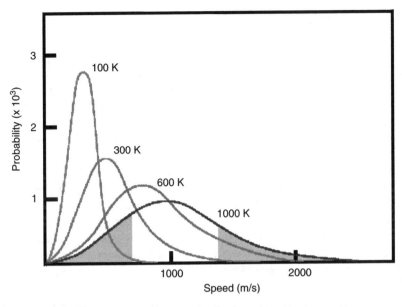

Figure 9.4 Maxwell–Boltzman distributions at various temperature

extremity of each spring there will be atoms with almost identical velocities, so the entire structure of the atomic system will be relatively unperturbed. If a high temperature is chosen (for instance 1000 K, red curve) then a high percentage of atoms will display very different velocities (areas shaded in grey under the red curve). So at the extremity of each spring are atoms with very different velocities and then the space occupied by the atomic structure will be increased, allowing exploration of its conformation.

Molecular Dynamics is relatively insensitive to temperature. Increasing the temperature will provide energy to overcome the rotational barriers. If the conformational space available to a system is the main area of investigation then a high temperature should be chosen (typically 1000 K or higher). Alternatively if the behaviour of a system at a particular temperature (eg the function of an enzyme at 37°C) is being studied then the simulation must be performed at this certain value.

9.4 Factors affecting the length of simulations

This is dependent on the time-scale of interest, as outlined below, and as a rule of thumb sampling times should be 10 times this value. For rotational barriers around 1 kcal/mol, the time for a rotation is ∼1.2 ps, hence the simulation should exceed 12 ps; for rotational barriers around 5 kcal/mol the time for a rotation is ∼1.5 ns, hence the simulation should exceed 15 ns; for rotational barriers around 10 kcal/mol the time for a rotation is larger than 1 ms and then simulation will be extremely difficult. It is important to note that the thermal agitation at 300 K is around 2.5 kcal/mol.

9.5   Problems caused by solvents

For small organic compounds the presence of a solvent is not really important. This is not the case with biological systems for which the solvent, often water, has a large influence on the global behaviour of the biomolecules. Even if implicit solvation is taken into account by the software, its use must be carefully considered because of the numerous possible variables and the fact that it will add a large number of atoms to the simulated system and hence will dramatically slow down the computational speed.

There are several methods used to examine solvation of a protein, but the main technique used in most software is the periodic boundary condition convention (PBC). In this convention the solvated protein is isolated in a virtual box as shown in Figure 9.5. Then this box is repeated in space in a $(3 \times 3 \times 3 = 27)$ box array, as shown in Figure 9.6 (only one of the three planes is shown for clarity). In this representation, it is necessary to calculate the interactions between the original box and all its images, which implies large computational time. The main advantage of this convention is to prevent evaporation of solvent: if a water molecule escapes from the central original box by the right side, then its image from the left box enters the central box by the left side (see Figure 9.7). These conditions allow a simulation of a system with a constant number of atoms.

In the case of lipid bilayer simulation, it is just necessary to build a bilayer from the left side to the right side of the box. With the PBC convention, the images of the bilayer included in the central box will ensure continuity through the different boxes (see Figure 9.8).

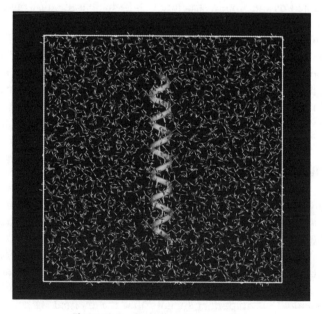

Figure 9.5   Periodic box convention

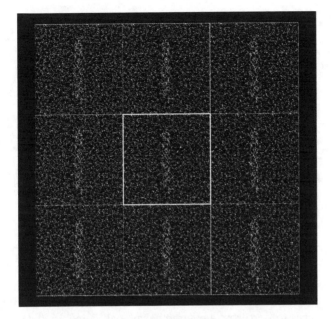

Figure 9.6 The final calculated system

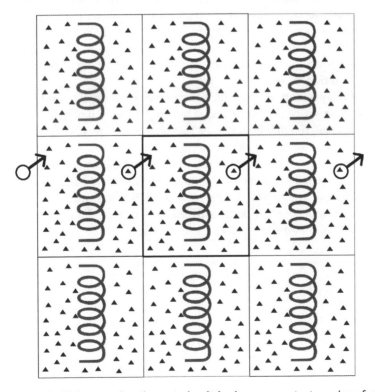

Figure 9.7 In PBC convention the central unit is always a constant number of atoms

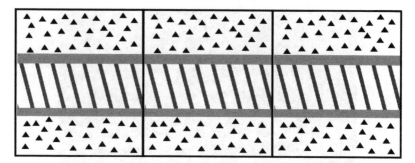

Figure 9.8   Schematic representation of a lipid bilayer building

9.6   How to build a lipid bilayer for simulation purposes

First some experimental parameters are needed (even if only approximations), like a specific surface per lipid (S) and the height of the bilayer (H). The simplest way to build a lipid layer is to start from a centred hexagonal lattice. In this system the basic unit is a losenge and hence an equilateral triangle is sufficient to build the whole system (Figure 9.9), equivalent to piling up lines of lipids shifted by half a period. From the specific surface per lipid (S) the distance between lipids (a) and the shift between lines (b) can be deduced: using very simple geometrical calculations $a = (S/2 * \sqrt{(3)})^{1/2}$ and $b = a * \cos(30)$. Note that if you build a lattice of m lipids on X and n lipids on Y, n must be even in order to ensure auto complementarity and there is no condition on m.

To build a bilayer starting from the above layer is straightforward by duplicating the layer, inverting it *vs* the z-axis and displacing it along this axis until the known height H is reached, roughly between the two phophorous planes. Then this assembly is put into the centre of a box with dimensions $(m.a) * (n.b) * c$ Å^3, with c being the height of the box. In order to ensure a full hydration of the lipid bilayer, c must be set in order to have around 20 molecules of water per lipid (at least). In these conditions, as full hydration of a phosphocholine moiety necessitates around 12 molecules of water, a noticeable amount of water will play the role of 'bulk' water. In standard conditions the height must be around 70 Å or more.

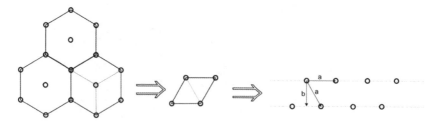

Figure 9.9   Building of a centred hexagonal lattice, simplification scheme

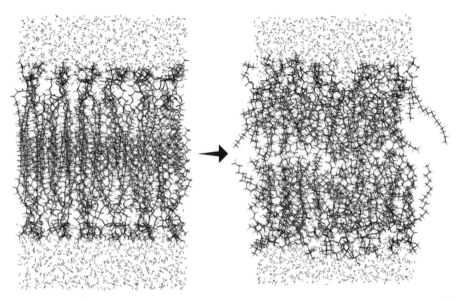

Figure 9.10 Model of 2 × 30 DMPC bilayer in water. On the left is the initial conformation of the starting bilayer with all-*trans* lipids; note the partly interdigitated bilayer. On the right is the same system after 2 ns at constant volume; note the presence of numerous gauche conformers and the absence of interdigitation

At this stage two layers constituted of individual lipids are being considered, generally in a crystalline conformation, arranged into a layer which has a surface corresponding to a liquid phase. As the lipids are in an all-*trans* conformation, the built layer is partially interdigitated. It is first necessary to 'melt' the alkyl chains: as this is a slow process it must be performed as the first part of the simulation at constant volume during at least 2 ns. During this time, gauche conformations appear and shorten the alkyl chains, giving slowly a correct bilayer (Figure 9.10). After this equilibration period, the productive run can be performed for at least 3 ns at constant pressure, producing accurate parameters during the last ns. For instance, interesting parameters which may be calculated are, for example, the lateral diffusion coefficient (Figure 9.11), solvation shells [see Figure 9.12(a,b)], radius of gyration of various complexes, orientation of interesting vectors (electronic dipoles, $P \ldots N$ vectors for phosphocholine headgroups, etc.)

It has to be remembered that a lipid exhibits very slow motion. Typically the lateral diffusion coefficient of a standard phosphocholine is around 10^{-8} cm^2/s (for water this coefficient is around 2.5×10^{-5} cm^2/s). In other words, a lipid undergoes a displacement roughly equivalent to 1 Å/ns. In a standard bio-membrane the average distance between lipids at the surface is around 7–8 Å, i.e., the time necessary to simply exchange the location of two neighbour lipids will be in the 10–100 ns range: (Figure 9.13).

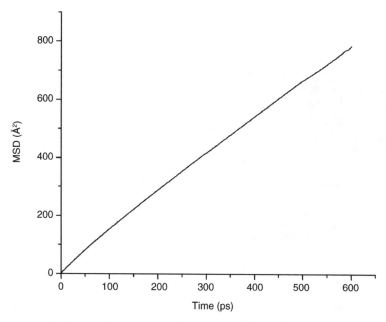

Figure 9.11   Plot of the mean square displacement (MSD, Å^2) of water molecules vs time (ps) for 0.6 ns of molecular dynamics at constant pressure. The slope of the curve is 1.2866. Following Einstein's equation, the lateral coefficient of water is one-sixth of this value, i.e., 0.214 Å^2/ps or 2.14×10^{-5} cm^2/s. In these conditions, the normal value for 'bulk' water is 3.5×10^{-5} cm^2/s. In fact a noticeable part of the water is partially trapped into hydration shells around charged moieties like phosphocholines

Other characteristic time parameters for standard lipids are the following:

- *trans*/gauche isomerization, 10–100 ps;
- head protrusion, 0.1–1 ns;
- rotation along main axis, 1–5 ns/turn;
- flipping between two layers, minutes to hours.

The very slow motion due to the lipids of a membrane model will have large implications on the various processes that can be studied by molecular dynamics. As a direct consequence, mixtures of lipids will prove to be extremely long in equilibrating. On the other hand, interaction or aggregation of peptides (or proteins) embedded in a lipid bilayer will be an extremely long process. In these conditions it is advisable first to attempt a study in an artificial interface, for instance, a slice of chloroform in water. This kind of system provides a very clean and smooth interface between hydrophilic and lipophilic media. Moreover, the chloroform molecules are small and spherical and exhibit a very high lateral diffusion coefficient. In these conditions interactions between proteins or peptides are largely accelerated. Nevertheless, one of the major drawbacks of this approach is the almost complete loss of lateral pressure exerted by a true lipid bilayer. This can lead to dramatic artefacts due to the instability of a large number of membrane helices in this artificial medium.

(a)
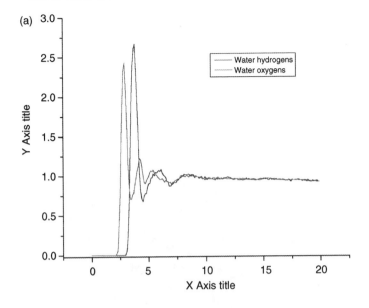

(b)
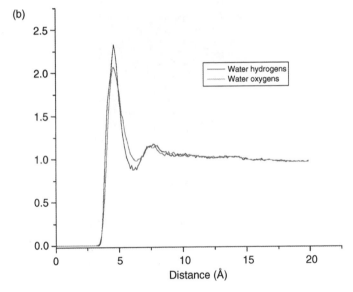

Figure 9.12 (a) The radial distribution function (RDF) for the phosphorus atoms of DMPC and oxygens (red) or hydrogens (blue) of water. Two hydration shells can be observed, the first at 4.6 Å, accounting for 3 H_2O, and the second at 6.8 Å, accounting for 9.5 H_2O. The first shell is characterized by a very sharp and narrow distribution. Moreover all oxygens are in the inner part of the sphere and all hydrogens in the outer part: the shell is perfectly orientated. This is characteristic of a hard ion like the phosphate moiety of the DMPC. This value is averaged on the 100 last frames. (b) the radial distribution function (RDF) for the ammonium group of DMPC and oxygens (red) or hydrogens (blue) of water. One unique hydration shell can be observed at 6.3 Å, accounting for 8.5 H_2O. The shell is characterized by a smooth and wide distribution. Moreover oxygens and hydrogens are not localized: the shell is totally unorientated. This is characteristic of a soft ion like the quaternary ammonium moiety of the DMPC. This value is averaged on the 100 last frames

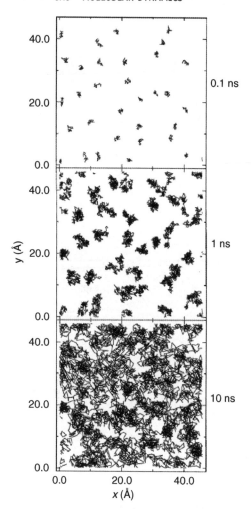

Figure 9.13 Stroboscopic view of a 64-DPPC bilayer. Only phosphorus atoms trajectories are shown according to elapsed time (copyright Essman and Berkowitz, 1999)

9.7 Special case of membrane proteins

In this case it is absolutely necessary to embed the protein in a coherent lipid bilayer model, along with all the necessary water and ions or counter ions. This results in a very large system and the atoms belonging to the environment of the protein account for six to seven times the atoms of the protein (remember that computational times are proportional to the square of the number of atoms). The PBC convention is used throughout, allowing in fact the creation of a quasi-infinite multilayered system (Figure 9.14).

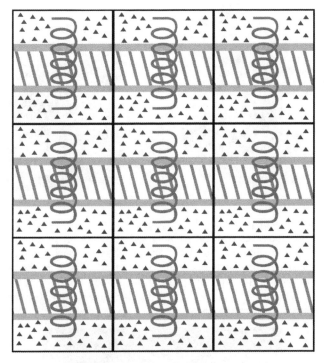

Figure 9.14 Schematic representation on the simulated system for a protein embedded within a membrane model (dark triangles represent water or ions)

The main problem with this kind of simulation is to embed the protein (or peptide) within the membrane model. This is not a trivial problem. One way is to start from the previously simulated bilayer and then to superimpose the protein onto the bilayer. *Note that the vertical position of the protein vs the polar interface of the membrane model is a crucial parameter.* It is necessary to withdraw lipids superimposed with or located at a too short distance from the protein. This threshold value is tricky to decide: it is not easy to adapt the complicated topology of the protein within a layer constituted of large molecules like phosphocholines. At this stage it is important to leave the smallest gaps possible between the protein and the bilayer even if some atoms are misplaced. One possible way to perform a full simulation is to first fix the backbone of the protein then perform a molecular dynamics run at constant volume during 1 or 2 ns in order to allow a full contact between the layer and the protein, then unfix the protein and perform a molecular dynamics run at constant volume during 1 or 2 ns in order to relax the full system (Figure 9.15) and then perform the productive run at constant volume or pressure as you prefer. This last run must be long, as the system consisting of a membrane protein embedded in a lipid bilayer is extremely sluggish.

Actually the largest calculations performed concern systems with about 300 000 atoms during a few ns. In specialized laboratories, systems between 100 000 and 25 000 atoms (Figure 9.16) are routinely simulated during 10–100 ns. Nevertheless, if

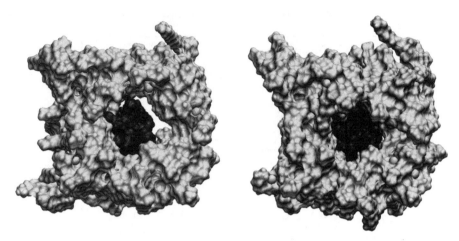

Figure 9.15 Simulation of a hydrophobic peptide embedded in a 2 × 30 DMPC bilayer. The peptide and lipids are shown as their accessibility surface (black for peptide and light grey for lipids); water molecules are not shown for clarity. On the left is the starting point of the simulation with the peptide embedded into the bilayer prior to any calculations: note the large gap between the peptide and the lipids in the upper right part of the peptide. On the right is the same system after 1 ns simulation at constant volume with peptide backbone fixed: note the almost complete contact between the peptide and the lipids

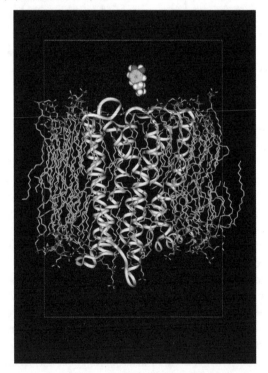

Figure 9.16 Simulation of the interaction of dopamine (in CPK spheres at the top of the box) with its own receptor (human D2 receptor of the GPCR'S superfamily, in a yellow ribbons) embedded in a POPC bilayer (pamitoyl oleyl phosphatidyl choline in green, blue, orange and red sticks). The whole system is surrounded by 3684 molecules of water (in red and white lines). The simulated box contains 23 109 atoms, including the counter ions necessary to ensure neutrality of the whole system

you consider the pH of the organism (7.4), the minimum system to simulate just in order to deal with a single proton is around 75×10^6 atoms, far exceeding the calculation power of the largest supercomputer.

Tips

If you want to know the number of atoms contained in any box, it is only necessary to calculate the volume of this box in Å^3 and divide this number by 10. This tip works well with water as a solvent. Note that in organic solvents the number of atoms will largely be less than this calculated value.

9.8 Summary

First of all, the use of molecular mechanics and dynamics pre-suppose a three-dimensional structure or a model built by sequence homology. Even if molecular mechanics involves large simplifications, it is an extremely powerful and reliable technique to study organic systems, except for reaction pathways. Molecular dynamics is a simple technique which is almost insensitive to temperature but extremely sensitive to the chosen time-step for the integration of the Newton's law. If the main interest of the user is to explore the conformational space of a system, then heating is the best choice (up to 1000 K or more).

With the most up-to-date supercomputers the space scale of the system will be, at most, $10 \times 10 \times 30$ nm (or equivalent) and the timescale at most 100 ns for systems in the $10 \times 10 \times 10$ nm space scale. In this case, calculations are huge and may imply several weeks or months of nonstop simulations. Apart from these drawbacks, there is almost no limitation to the size or complexity of the simulated systems. However, it has always to be remembered that the behaviour of biological systems is extremely slow at an atomic scale and will always necessitate huge amounts of calculation.

At present, starting from a protein one-dimensional sequence in order to build a three-dimensional structure is hopeless even for very short sequences. The best way to obtain access to a three-dimensional model is to perform sequence homology building, starting from an X-ray or NMR structure of a similar protein.

Further reading

Allen, M.P. and Tildesley, D.J. (1989). *Computer Simulation of Liquids.* Clarendon Press: Oxford.

Becker, O.M., MacKerell Jr, A.D., Roux, B. and Watanabe, M. (2001). *Computational Biochemistry and Biophysics.* Marcel Dekker: New York.

Frenkel, D. and Smit, B. (1996). *Understanding Molecular Simulations. From Algorithms to Applications.* Academic Press: San Diego, CA.

Leach, A.R. (2001). *Molecular Modelling. Principles and Applications*, 2nd edn. Prentice Hall–Pearson Education: Old Tappan, NJ.

Rapaport, D.C. (1995). *The Art of Molecular Simulation*, 2nd edn. Cambridge University Press: Cambridge.

Webster, D.M. (2000). *Protein Structure Prediction. Methods and Protocols*. Methods in Molecular Biology, Vol. **143**. Humana Press, Totowa, NJ.

10

Two-dimensional infrared studies of biomolecules

Xabier Coto, Ibón Iloro and **José Luis R. Arrondo**

Unidad de Biofisica (CSIC-UPV/EHU) and Departamento de Bioquìmica, Universidad del Pais Vasco, PO Box 644, E-48080 Bilbao, Spain

10.1   Introduction

The unique three-dimensional structure of a protein gives rise to a folding pattern that characterizes its specific biological function. The techniques that provide detailed information on protein three-dimensional structures are X-ray crystallography and nuclear magnetic resonance (NMR). These methods are limited by either the requirement of protein crystals or the size of the protein. Moreover, obtaining the structure of a protein does not immediately lead to understanding its molecular mechanism. Therefore, there is ample scope for aspects of protein structure and function to be elucidated by complementary, lower-resolution techniques, even in proteins with known three-dimensional structures. Infrared spectroscopy has been found to be a suitable technique for the study of protein conformation. In 1967 the observed amide I band from simple molecules dissolved in H_2O and D_2O was extended to characterize proteins. However, this work was limited to a qualitative approach of the secondary structure. The complexity of the amide I band and the presence of water (a strong infrared absorber) made it difficult to extract information from the infrared spectrum. The appearance of instruments based on interferometric devices with a high signal-to-noise ratio (known as Fourier-transform infrared spectrometers) and the application of data treatment techniques improved the resolution of the spectra to obtain what are termed in IR spectroscopy 'amide bands'. Specific components of the spectra are finally assigned to protein or subunit structural features. A different approach consists of the use of infrared spectroscopy

Chemical Biology   Edited by Banafshé Larijani, Colin. A. Rosser and Rudiger Woscholski
© 2006 John Wiley & Sons, Ltd

to examine ligand-induced changes produced in proteins during their functional cycles. The strategy used is called 'difference infrared spectroscopy' and consists of looking closely at the bands corresponding to protein or ligand groups that undergo changes during the transition from one state to another. By these means, it is possible to detect changes in a single amino acid residue. External perturbations such as temperature are commonly used to obtain a greater insight in protein structure by using infrared spectroscopy. This has been used to study conformational changes in proteins during unfolding induced by temperature.

More recently, Noda has proposed the use of infrared two-dimensional correlation spectroscopy (2D-IR) to increase the information that can be extracted from a spectrum. This approach, essentially different from 2D-NMR spectroscopy, uses correlation analysis of the dynamic fluctuations caused by an external perturbation to enhance spectral resolution without assuming any line shape model for the bands. The technique was intended for the study of polymers and liquid crystals, and it has recently been applied to proteins. In the latter case, the perturbation can be achieved through changes in temperature, pH, ligand concentration and lipid-to-protein ratio.

Infrared spectroscopy can be used in turbid suspensions, such as membranes or with big proteins, but the methods for studying the spectrum are impaired by the difficult interpretation of the composite bands obtained from proteins. Thus, more powerful methods of spectral analysis are needed

---

10.2   Description of the technique

The use of the two-dimensional (2D) correlation approach produces an increase in the spectral resolution. The effect of a perturbation such as a change in temperature can be followed and from the response of the system to the perturbation, a dynamic spectrum is obtained so that physical information, such as a conformational change can be distinguished.

10.3   Spectral simulations

Most studies using 2D-IR have interpreted the synchronous and asynchronous maps in terms of changes in intensity. However, shifts in band position, which can be associated with protein environmental changes, or changes in bandwidth can affect the maps, mainly the asynchronous one. Therefore, spectral simulations are needed to understand the changes observed. Gaussian peaks are generated using the general Gaussian equation:

$$P(x) = \frac{1}{\sigma\sqrt{2\pi}} e^{-(x-\mu)^2/(2\sigma^2)} \qquad (10.2)$$

The power of the 2D correlation approach results primarily in an increase of the spectral resolution by dispersal of the peaks along a second dimension that also reveals the time-course of the events induced by the perturbation. Correlations between bands are found through the so-called synchronous (Φ) and asynchronous (Θ) spectra that correspond to the real and imaginary parts of the cross-correlation of spectral intensity at two wave numbers, i.e. two vibrations of the protein characterized by two different wave numbers ($\nu1$ and $\nu2$) are being affected at the same time (synchronous) or the vibrations of the functional groups corresponding to the different wave numbers change each at a different time (asynchronous).

In a synchronous 2D map, the peaks located in the diagonal (autopeaks) correspond to changes in intensity induced in our case by temperature, and are always positive. The cross-correlation peaks indicate an in-phase relationship between the two bands involved.

Asynchronous maps show not-in-phase cross-correlation between the bands and this gives an idea of the time-course of the events produced by the perturbation.

To obtain two-dimensional infrared correlation spectra the formalism proposed by Noda is usually followed. From the response of the system to the perturbation, a dynamic spectrum is obtained so that physical information can be found. Once the dynamic spectrum has been obtained as a matrix formed by the spectra ordered according to the change produced by the perturbation, the Fourier transform gives two components, the real one corresponding to the synchronous spectrum and the imaginary one corresponding to the asynchronous one (Figure 10.1). In terms of calculation, instead of using a Fourier transform that would require large computation times, an adequate numerical evaluation of 2D correlation intensity is used.

Thus, the synchronous 2D correlation intensity can be expressed as

$$\Phi(\nu_1, \nu_2) = \frac{1}{m-1} \sum_{j=1}^{m} \bar{y}_j(\nu_1) \cdot \bar{y}_j(\nu_2) \tag{10.1}$$

$\bar{y}_j(\nu_i)$ is the dynamic spectra calculated from the spectral intensities as a deviation from a reference spectrum at a point of physical variable t_j.

The computation of asynchronous 2D correlation intensity is somewhat more complicated. Two approaches can be used: (i) using the Hilbert transform and (ii) a direct procedure, obtaining similar results (for a detailed discussion on the asynchronous calculation, see the Further Reading section)

In all simulations each dynamic spectrum consisted of 13 spectra with 130 points equivalent to an amide I band in the region 1700–1600 cm^{-1} with a nominal resolution of 2 cm^{-1}. The changes were assumed to be linear. The maps were obtained assuming that one or two peaks were changing.

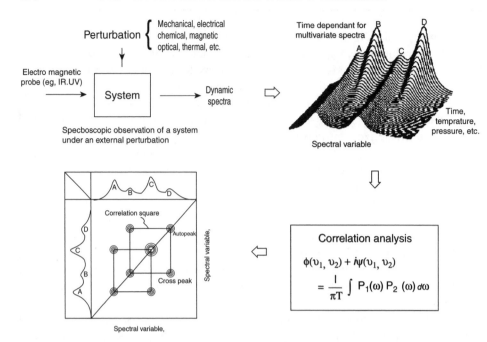

Figure 10.1 General scheme for obtaining 2D-IR spectra. From the response of the system to perturbation, a dynamic spectrum is obtained. The Fourier transform gives two components, the real one corresponding to the synchronous spectrum and the imaginary one corresponding to the asynchronous one

---

Spectral simulations are of great help in interpreting spectra because they give information in a noise-free environment and, since the variations of the different parameters can be controlled, their effect on the spectra can be gradually studied.

10.3.1   Intensity changes

Changes in intensity were observed in spectra composed of one or two peaks. The corresponding correlation maps are represented in Figure 10.2. The bottom and top maps correspond, respectively, to spectra generated from changes in one or two peaks. The left-hand traces correspond to synchronous spectra while the right-hand traces are from asynchronous spectra. Negative peaks are represented as shaded areas. Asynchronous spectra consist only of noise when only the band intensity has been changed. Similar results are obtained in simulations of 2D electronic correlation

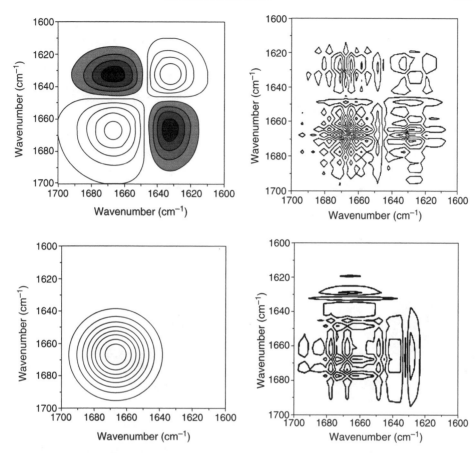

Figure 10.2 Bidimensional correlation maps corresponding to changes in intensity of a band composed of one (bottom) or two peaks (top). Synchronous maps are located at the left and asynchromous to the right. The x- and y-axes correspond to the numbers of the point on the artificial curve and the z-axis is the correlational intensity. Roughly they represent a protein amide I band in a D_2O buffer. Negative peaks are shaded

spectra or using Lorentzian band shapes. The lack of bands in the asynchronous spectra has been observed in protein thermal denaturation after aggregation has taken place, and this is not a noise-artefact due to the decrease in spectral intensity after aggregation, because artificial spectra are noise-free and yet the asynchronous map consists only of low-amplitude noise. Thus, the lack of bands is a characteristic of the denatured proteins once the aggregation has been completed. On the other hand, considering that synchronous spectra reflect in-phase changes, differences in peak intensity are expected to produce changes. It can be seen that the complexity of the synchronous spectra changes as a function of the number of peaks involved. If only one peak is changing, one autopeak is seen. If two or more peaks are involved, the respective cross-peaks show the way in which intensities are changing. If both increase or decrease simultaneously, the cross-peaks are positive, and negative if one peak is increasing and the other one is decreasing.

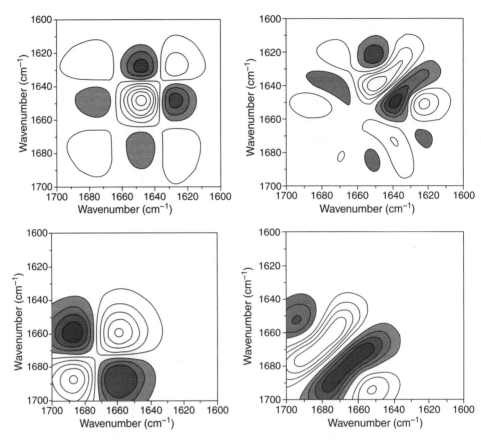

Figure 10.3   Bidimensional correlation maps corresponding to changes in position (band shifting) of a band composed of one (bottom) or two peaks (top). Synchronous maps are located on the left and asynchronous on the right. Axes are as in Figure 10.2

10.3.2   Band shifting

Changes in band position were analysed in a similar way. Simulated bands were constructed with one or two components changing simultaneously, and the synchronous and asynchronous maps are represented in Figure 10.3. In all the synchronous maps, autopeaks are located at the initial and final positions of the maxima during band shifting. This is the reason why two peaks are observed in the synchronous map corresponding to a change of band position in one peak. Cross-peaks are positioned next to the equivalent autopeaks with a negative intensity since the intensity at a given frequency at the beginning of the shift decreases, whereas intensity at a different frequency will increase after the shift. The characteristic pattern for a band shift is observed in the asynchronous spectra as a butterfly or banana shape, depending on whether we consider the full symmetric map, or only one side of the diagonal. These peaks include the frequencies through which the maximum has moved throughout the band shifting. As in the synchronous spectra, there are also cross-peaks of the opposite sign between the beginning and the end of the 'bananas'.

10.3.3   Bandwidth

The same approach as in the two previous models was used for band broadening. In this case the pattern obtained is more complex. It has to be taken into account that changes in bandwidth in proteins are not usually a result of a phase transition as in the case of lipids, but they are rather the consequence of changes in conformation and subsequent variation in the area of the band associated with that conformation. Hence, changes in bandwidth can be clearly discerned because a more complicated correlation map is produced, but they are not easily assigned (see below). In the case of the models studied, the resulting synchronous maps (Figure 10.4) show patterns with several autopeaks located at the peak inflection points, since these are the

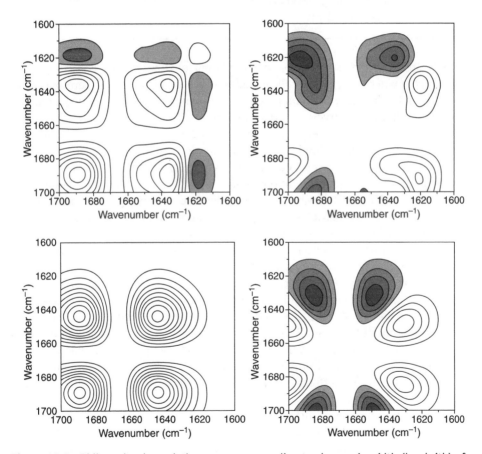

Figure 10.4   Bidimensional correlation maps corresponding to changes in width (bandwith) of a band composed of one (bottom) or two peaks (top). Synchronous maps are located on the left and asynchronous on the right. Axes are as in Figure 10.2

frequencies at which greater intensity variations occur throughout the band broadening. These autopeaks are associated with positive cross-peaks because both points are changing their intensity in the same direction. The asynchronous spectrum is much more complicated, showing an increased number of peaks with different sizes and intensities.

These simulations show that other effects than true asynchronisms, such as band shifts or bandwidth changes, can generate apparent asynchronous peaks that can be misinterpreted, especially in highly overlapping peaks

10.4   Two-dimensional studies of human lipoproteins

Human lipoproteins are a very interesting example to apply two-dimensional spectroscopy, because they are composed of lipids and proteins, like membranes, and they have large proteins, like apoB100, which is a single polypeptide of 4536 aminoacyl residues.

Lipids are transported in human blood plasma by lipoproteins consisting of a nonpolar core where triacylglycerols and cholesteryl esters are hidden surrounded by a monolayer facing the water composed of phospholipid, cholesterol and proteins, giving these lipid-rich structures water solubility. Blood plasma lipoproteins are classified on the basis of their density, which in turn is a reflection of their lipid content. The greater their lipid contents the lower their density. There are three different classes used in infrared studies. These are VLDL (very-low-density lipoprotein), LDL (low-density lipoprotein) and HDL (high-density lipoprotein). Biosynthesis of VLDL cholesterol, triglycerides and apolipoproteins takes part in liver hepatocites. These particles have apoB-100, apoC and apoE proteins. VLDL is converted into LDL, cholesteryl ester-rich particles that have a single molecule of apoB-100. LDL carries cholesterol from liver to peripheral tissues where it is used in membrane and steroid biosynthesis. HDL is synthesized from liver precursors and is matured in plasma. Apolipoproteins in HDL are apoA-I and apoA-II. HDL removes cholesterol from the cells carrying it to the liver in a 'reverse transport'. The amide I bands of VLDL, LDL and HDL are shown in Figure 10.5. The major apolipoprotein in VLDL is apoB-100, which is the single protein molecule in LDL. The spectrum of this protein is very characteristic and dominates the amide I. On the other hand, the HDL spectrum is similar to many other helical proteins. It is interesting to notice the different spectrum obtained can be related to the different function of the lipoproteins.

One interesting feature of infrared spectroscopy is that in the same spectrum the amide bands corresponding to the protein and the bands corresponding to the lipid can be studied. The core-associated lipids have been shown to undergo an order/disorder transition near human body temperature, the actual transition temperature being determined by the lipid composition. Figure 10.6 shows the synchronous (left)

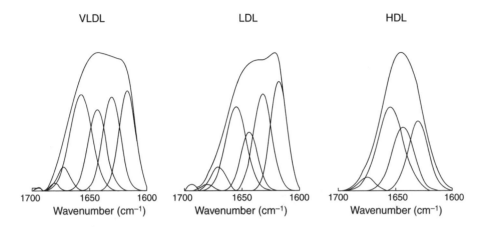

Figure 10.5   Amide I bands and their components corresponding to VLDL (left), LDL (middle) and HDL (right)

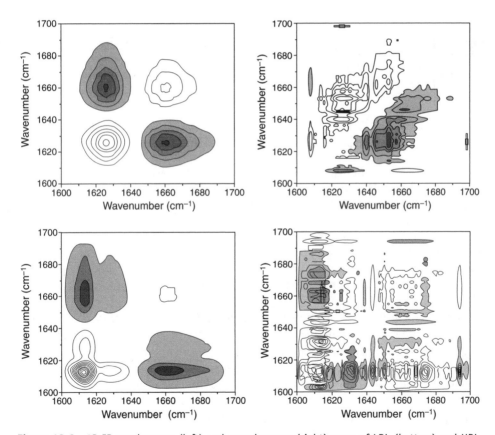

Figure 10.6   2D-IR synchronous (left) and asynchronous (right) maps of LDL (bottom) and HDL (top) in the 1600–1700 cm^{-1} region corresponding to the amide I region at the 20–40 °C interval. The spectra were taken in a D$_2$O medium

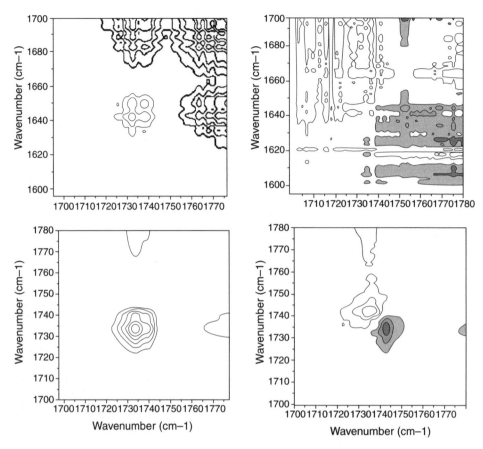

Figure 10.7 Synchronous (left) and asynchronous (right) 2D-IR maps of LDL (bottom) and HDL (top) in the 1700–1800 cm^{-1} region corresponding to the lipid carbonyl region at the 20–40 °C interval. The spectra were taken in a D$_2$O medium

and asynchronous (right) maps of LDL (bottom) and HDL (top) corresponding to the spectra obtained after heating the particles from 20 to 40 °C in a D$_2$O medium. The same maps in the region 1700–1780 cm^{-1} corresponding to the C=O stretching vibrational model corresponding to the lipids are shown in Figure 10.7.

In the synchronous map corresponding to LDL, two autopeaks at 1617 and 1660 cm^{-1} are observed with the corresponding cross-peaks being negative, indicating that one is increasing whereas the other is decreasing in intensity. From classical infrared studies, an increase in α-helix content had been proposed, which was confirmed by the two-dimensional studies. In addition, the asynchronous map shows only noise, which was shown in Section 10.3.1 with the simulation maps that correspond only to changes in intensity. In HDL, two autopeaks and their corresponding negative cross-peaks are also observed, but in this case they are located at 1626 and 1660 cm^{-1}, which indicates a different pattern in the protein than in LDL.

In the maps corresponding to the lipid (Figure 10.7) a clear difference can be seen between LDL and HDL. Whereas in LDL an autopeak is seen in the synchronous map at $1636\,cm^{-1}$ and a cross-peak 1736/1742 (+), in HDL only noise is produced. As shown earlier, (Section 10.3), in LDL one peak is changing, but not only in intensity. Infrared studies of phospholipids have shown the presence of two bands corresponding to two different hydration states of the interface carbonyl. The asynchronous map would point to a change in the lipid core where different hydration states could be involved. In the case of HDL, no lipid transition was proposed, and the 2D-IR agrees with the absence of conformational changes in the lipid in the interval 20–40 °C.

10.5 Summary

The use of infrared spectroscopy increases knowledge of protein structure as a complement or an alternative to other techniques. The full potential of the technique has not been developed due to some setbacks. Two-dimensional infrared spectroscopy can be of great help in extracting more information from the spectra. Two-dimensional infrared correlation maps obtained from artificial bands with known changes in band parameters give rise to characteristic patterns. Biological systems of varied complexity will give rise to 2D-IR correlation maps that reflect the changes induced by the perturbation and can be interpreted as a composite of the band simulation maps. The use in complex particles such as lipoproteins, with a lipid and a protein moiety, is a clear example of the potential of 2D-IR and how simulations help in the interpretation of the correlation maps, both in the protein and the lipid bands, distinguishing between changes only in intensity, or changes in band position or bandwidth.

Further reading

Arrondo, J.L.R. and Goñi, F. M. (1997). Infrared spectroscopic studies of membrane lipids. In *Biomolecular Structure and Dynamics*, Vergoten, G. and Theophanides, T. eds, pp. 229–242. Kluwer Academic: Dordrecht.

Arrondo, J.L.R. and Goñi, F. M. (1999). Structure and dynamics of membrane proteins as studied by infrared spectroscopy. *Prog. Biophys. Mol. Biol.* **72**: 367–405.

Arrondo, J.L.R., Iloro, I., Aguirre, J. and Goñi, F.M. (2004). A two-dimensional IR spectroscopic (2D-IR) simulation of protein conformational changes. *Spectroscopy* **18**: 49–58.

Bañuelos, S., Arrondo, J. L. R., Goñi, F. M. and Pifat, G. (1995). Surface–core relationships in human low density lipoprotein as studied by infrared spectroscopy. *J. Biol. Chem.* **270**: 9192–9196.

Barth, A. and Zscherp, C. (2002). What vibrations tell us about proteins. *Q. Rev. Biophys.* **35**: 369–430.

Coto, X. and Arrondo, J.L.R. (2001). *Infrared Spectroscopy of Lipoproteins*, Pifat-Mrzljak, G., ed., pp. 75–87 Kluwer Academic: New York.

Lefevre, T., Arseneault, K. and Pezolet, M. (2004). Study of protein aggregation using two-dimensional correlation infrared spectroscopy and spectral simulations. *Biopolymers* **73**: 705–715.

Noda, I., Dowrey, A.E., Marcott, C., Story, G.M. and Ozaki, Y. (2000). Generalized two-dimensional correlation spectroscopy. *Appl. Spectrosc.* **54**: 236a–248a.

Zscherp, C. and Barth, A. (2001). Reaction-induced infrared difference spectroscopy for the study of protein reaction mechanisms. *Biochemistry* **40**: 1875–1883.

11

Biological applications of single- and two-photon fluorescence

Banafshe Larijani[1] and Angus Bain[2]

[1]*Cell Biophysics Laboratory, Cancer Research UK, London Research Institute, 44 Lincoln's Inn Fields, London WC2A 3PX, UK*
[2]*Department of Physics and Astronomy and CoMPLEX, University College London, Gower Street, London WC1E 6BT, UK*

11.1   Introduction

The main objective of this chapter is to highlight how single- and two-photon fluorescence may be exploited in biology, notably in the imaging of molecular interactions within cells. Single-photon fluorescence has been exhaustively reviewed in a number of textbooks (see Further reading) and will not be duplicated here. There are however, significant differences between single- and two-photon excitation which merit consideration as these underpin the growing importance of two-photon fluorescence as a fundamental tool in biological imaging. Moreover two-photon fluorescence provides different and additional molecular information. These concepts are discussed in terms of spatial resolution in fluorescence microscopy and fluorescent probe measurements with detailed biological applications provided by single- and two-photon fluorescence lifetime imaging of resonance energy transfer (RET) in intracellular signalling.

11.2   Basic principles of fluorescence

Fluorescence is one of the two types of luminescence. It is the emission that results from the transition of paired electrons in a higher energy state to a lower energy state. Such

Chemical Biology   Edited by Banafshé Larijani, Colin. A. Rosser and Rudiger Woscholski
© 2006 John Wiley & Sons, Ltd

transitions are 'quantum mechanically' allowed and their lifetimes are near 10 ns. Once a molecule absorbs energy it is excited to a higher energy level. One of the ways that it can return to the ground state is by emission of fluorescence. Substances that exhibit fluorescence generally possess delocalized electrons that are present in conjugated double bonds. These compounds are known either as chromophores or fluorphores.

Fluorescence spectral data are generally presented as emission spectra. The 'smoothness' of the spectra of different compounds is dependent on the individual vibrational energy levels of the ground and excited states. Those with a vibrational structure have prominent shoulders on their emission spectra and those devoid of vibrational structures show a smooth Gaussian curve. Two general characteristics of fluorescence emission are Stokes' shift, where a loss of energy is observed between excitation and emission, and the mirror image phenomenon, where the fluorescence emission spectrum is the mirror image of the absorption spectrum.

Fluorescence has been exploited as a tool for investigating structure and the dynamics of molecules in their microenvironment. The reason behind this is that fluorescent molecules are highly sensitive to their surroundings. Therefore the emission of fluorescence is sensitive to physical and chemical parameters such as pH, pressure, viscosity, electric potential, quenchers, temperature and ions and can be used as a reporter of molecular environment.

There are many photophysical processes that are responsible for the de-excitation of molecules. The few examples of intermolecular photophysical processes that induce fluorescence quenching are electron transfer, proton transfer and energy transfer. The following section will focus on energy transfer and most specifically on nonradiative energy transfer.

11.3   Main principles of RET via single-photon excitation

Resonance energy transfer is the physical process by which energy is transferred nonradiatively from an excited molecular chromophore (the donor D) to another chromophore (the acceptor A) by means of intermolecular long range dipole–dipole coupling. The energy transfer is nonradiative; that is D does not actually emit a photon and A does not absorb a photon. The term 'resonance energy transfer' is specifically used as opposed to fluorescence resonance energy transfer (FRET) since it is not the fluorescence that is being transferred but the electronic energy of the donor D to the acceptor A. This energy transfer can occur via different interaction mechanisms. They can be either long-range interactions that result from Coulombic interactions (80–100 Å) or they can be short range interactions (<10 Å) that occur due to multipolar interactions.

In order for RET to occur over distances of 10–100 Å the fluorescence spectrum of D and the absorbance spectrum of A have to overlap adequately; the quantum yield of D and the absorption coefficient of A (ε_A) must be sufficient. Furthermore, for the dipole–dipole interaction to occur, the transition dipoles of D and A must either be oriented favorably relative to each other or one or both of them have to have a certain degree of rapid rotational freedom.

Förster in 1946, described this resonance energy transfer by the following equation:

$$\kappa_T = 1/\tau_D[(Ro/r)^6]$$

where τ_D is the fluorescence lifetime of the donor; r is the distance between D and A; and Ro is the Forester critical radius and that is calculated from the spectroscopic and mutual dipole orientational parameters of D and A. The process of energy transfer can be illustrated as a transition between two states where κ_T is the rate of transfer between the two states i.e.

$$[D^*, A]\kappa_T \rightarrow [D, A^*] \quad (* \text{ is the excited state})$$

11.4 Detection of RET

There are two modes of detection; one is in steady-state and the other is time-resolved.

11.4.1 Steady-state method

Donor quenching is the most common method used for detecting energy transfer. Quenching of the donor fluorescence is due to acceptors at different distances as well as orientations, and due to the motions of donor or acceptor. Excitation is set at the wavelength of donor absorption and the emission of the donor is monitored. The emission wavelength of donor is selected such that no contribution from the acceptor fluorescence is observed.

Another method for detecting energy transfer is by monitoring the fluorescence intensity of the acceptor. The fluorescence intensity of the acceptor is enhanced when energy transfer occurs (when the donor is excited). In an emission spectrum, one excites at the wavelength of donor absorption and observes the intensity increase of the acceptor. In an excitation spectrum, one sets detection at the acceptor emission wavelength and observes enhancement of intensity at wavelength range where donor absorbs.

As can be seen from the Förster equation, the dimensions of the system under study are the most important factor to be considered when choosing a D and A system. The choice of the donor/acceptor molecule is dictated by several factors:

(1) The excitation spectrum of the acceptor should have a maximum overlap with the emission spectrum of the donor.

(2) The maximum absorption of the donor should coincide with the maximum output of the light source.

(3) There should be a low quantity of direct excitation of the acceptor at the excitation maximum of the donor.

(4) Minimization of self-transfer – the donor has to have a small overlap between its own emission and absorption spectrum.

   The process of RET results in a decrease in the maximum intensity (I_D) of the donor emission (quenching) and a simultaneous increase in the maximum intensity (I_A) of the acceptor emission. Consequently the ratio of $I_A : I_D$ can be taken as the measurement of RET. The value of this ratio depends on the average distance between the donor and the acceptor pairs. An increase in the average distance between the donor and the acceptor chromophores results in the $I_A : I_D$ ratio decreasing and reaching a certain constant value. However before using this $I_A : I_D$ as an experimental measurement of RET, the proper donor–acceptor ratios have to be determined by varying the donor–acceptor ratios over a wide range.

   Artefacts such as self-absorption, collisional quenching and quenching by formation of a complex that has a low quantum yield can affect the occurrence and detection of RET. Collisional quenching affects the lifetime of the excited state and the second type of quenching reduces the concentration of potentially fluorescent molecules.

   RET can be distinguished from re-absorption complex formation and collisional quenching by verifying that the change in I_A is independent of the sample volume and viscosity and that there is no change in the donor fluorescence or its absorption spectrum.

11.4.2 Time-resolved method

Steady-state and time-resolved measurements are related to one another by the following relation:

$$I_{(t)} = I_0 \, x \, e^{t/\tau}$$

where τ is the single decay time of a specific chromophore, and I_0 is the intensity at time $t = 0$, immediately after the excitation pulse.

   In this method the fluorescence decay of the donor after pulse excitation is measured. Therefore the lifetime of the photon emitted upon fluorescence can be measured. This lifetime is the intrinsic property of the donor chromophore and its value is independent of variations in concentration. This fact is a major advantage of time-resolved methods over steady-state methods. If the decay is a single exponential, the measurement of the decay time in the presence and absence of the acceptor is straightforward. That means that, in circumstances where the decay is single exponential and the donor lifetime changes in the presence of the acceptor, then transfer of energy has taken place. The efficiency of this energy transfer can be calculated by:

$$\text{RET efficiency} = 1 - \langle \tau_{D/A} \rangle / \langle \tau_D \rangle$$

where $\langle \tau_{D/A} \rangle$ is the average decay time of the donor in the presence of the acceptor and $\langle \tau_D \rangle$ is the average decay time of the donor alone.

However, if the decay is not a single exponential in the absence of the acceptor, then this may be due to the heterogeneity of the microenvironment of the donor chromophore. Moreover the donor–acceptor distance can also be calculated from the following relation:

$$r = R_0/[(\tau_D/\tau_{D/A}) - 1]^{1/6}$$

Owing to the sixth power dependence the change in the transfer efficiency is sharp when $r = R_0$. The transfer of energy is close to 1 when $r < R_0/2$ and it approaches 0 when $r > R_0/2$. Thus the distance between the donor and the acceptor is in the following range:

$$0.5\,R_0 < r < 1.5\,R_0$$

To enable the measurements of lifetime decays of an excited state chromophore within a single cell, fluorescence lifetime imaging microscopy (FLIM) can be used. Imaging the change in lifetime of a chromophore in a cellular environment provides information on the local physical and chemical properties of the cellular environment. FLIM is a powerful tool that can be exploited both in fixed and live cells. It can be used to quantify RET between two species labelled with two different chromophores. FLIM is either in the time domain FLIM or in the frequency domain. Both of these types of instruments can be utilized to measure lifetime variations in single cells. The main differences between these methods will be depicted here. However, to have a wider understanding of FLIM it is essential to refer to the suggested references. Briefly, time domain FLIM is feasible by the combination of single-photon timing techniques and scanning techniques. In the case of time-domain FLIM the light source is a pulsed laser. The lifetime can be directly detected by time-domain, and its major asset is that multiexponential decays can be directly monitored. However, the relatively long measurement time needed for collecting photons at each point can be problematic if, in live cells, short reaction times need to be measured.

With frequency domain FLIM the light source is a continuous wave laser as opposed to a pulsed laser. The continuous wave laser is modulated via an acousto-optical modulator and the sample is excited by a sinusoidally modulated light. The fluorescence response is also sinusoidally modulated at the same frequency but it is delayed in phase and is partially demodulated. For a single exponential decay the lifetime of the donor chromophore can be quickly calculated by either the phase shift ϕ (τ_p) or the modulation ratio M (τ_m) using the following equations:

$$\tau_p = 1/\omega \tan^{-1} \phi$$
$$\tau_m = 1/\omega(1/M^2 - 1)^{1/2}$$

The above relations show how the phase shift and the relative modulation are related to the decay time. Although single frequency-domain FLIM has the advantage of being very rapid, it is not the best tool to directly measure multiple exponential decays. Nevertheless multiple frequency FLIM instruments are used for detecting multiple exponential decays that may arise due to heterogeneity of the local environment of the chromophore.

11.5   Biological examples of RET monitored by frequency-domain FLIM

RET in the last decade has been adopted from its sole usage in chemistry as a 'spectroscopic ruler' to be applied in the life sciences as a 'biological ruler'. To observe the physiological behaviour of signalling molecules it has been crucial to monitor their behaviour within the cellular environment without major perturbations. Today RET is used in microscopic imaging as a tool to measure in a precise manner inter- and intra-molecular distances in live or fixed cells.

During cellular signal transduction, proteins continuously sense various membrane compartments and associate with specific proteins or lipids in response to localized physiochemical changes. Fluorescence imaging methods have been utilized to quantify and localize protein–protein and protein–lipid associations in cells.

Using RET detected by frequency-domain FLIM, Larijani and co-workers have shown the regulated interaction of a lipid transfer protein–phosphatidylinositol transfer protein (PITPα) with phosphatidylcholine (PtdCho) and phosphatidylinositol (PtdIns) at the plasma membrane of COS-7 cells. By exploiting RET via decrease in the lifetime of the donor molecule molecular, associations can be reported. They have illustrated that the membrane association of PITP upon growth factor stimulation is a regulated phenomenon. Using this method they have illustrated with various Ras/ MAP kinase inhibitors that the depletion of PtdIns from the membrane is not the sole driving force required for the recruitment of PITPα to the membrane. These types of observations would have gone unnoticed if precise tools such as RET measured by FLIM had not been exploited.

Figure 11.1 depicts the effect of growth factor (EGF) stimulation on PITPα association with the plasma membrane of COS-7 cells. The GFP-PITPα (donor) lifetime is reduced from 2 to 1.7 and 1.6ns in the presence of the acceptors BODIPY-PtdCho and BODIPY-PtdIns, respectively. This localized change in GFP lifetime is only observed at the plasma membrane upon stimulation with EGF.

The second example illustrates how enhanced acceptor fluorescence-resonance energy transfer (EAF-RET) can be exploited to monitor the conformation change of an intact serine/threonine kinase, PKB/Akt, upon growth factor stimulation (PDGF) in NIH3T3 cells. Such kinases are responsible for regulating activation and deactivation of cell signalling. These fundamental biological processes depend on conformational changes of these proteins. This unique method of analysis (EAF-RET) has permitted Calleja and co-workers to observe the change in conformation of PKB/Akt, at the plasma membrane where it adopts an open conformation due to interaction with phosphoinositides such as PtdIns $(3,4)P_2$ and PtdIns $(3,4,5)P_3$ (Figure 11.2). Once

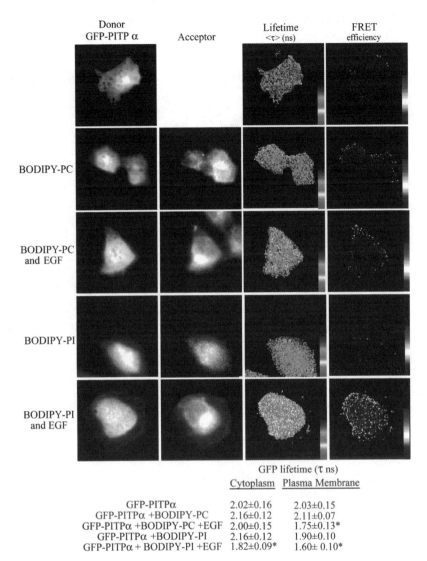

	Donor GFP-PITP α	Acceptor	Lifetime ⟨τ⟩ (ns)	FRET efficiency

	GFP lifetime (τ ns)	
	Cytoplasm	Plasma Membrane
GFP-PITPα	2.02±0.16	2.03±0.15
GFP-PITPα +BODIPY-PC	2.16±0.12	2.11±0.07
GFP-PITPα +BODIPY-PC +EGF	2.00±0.15	1.75±0.13*
GFP-PITPα +BODIPY-PI	2.16±0.12	1.90±0.10
GFP-PITPα + BODIPY-PI +EGF	1.82±0.09*	1.60± 0.10*

Figure 11.1   EGF-stimulated interaction of EGFP-PITPα with BODIPY-PC and BODIPY-PI monitored by fluorescence lifetime imaging in intact cells. (a) EGFP-PITPα transfected COS-7 cells were incubated with BODIPY-PC or BODIPY-PI and stimulated with EGF and fixed. RET results in the shortening of the EGFP (donor) fluorescence lifetime, which is measured by two parameters, the phase shift (τ_p) and relative modulation (τ_m). The average lifetime $\langle \tau \rangle = ([\tau_p + \tau_m]/2)$ of EGFP-PITPa (donor) in cells that are only incubated with BODIPY-PC and BODIPY-PI (acceptors) does not vary significantly. In the presence of EGF, there is a significant decrease in the lifetime of EGFP-PITPa at the plasma membrane (at 95% confidence interval $p < 0.0001$, $n = 10$). The EGFP lifetime pseudocolor scale is from 1.5 ns (red) to 2.1 ns (dark blue). FRET efficiency is calculated as $E = 1 - \tau_{da}/\tau_d$, where τ_{da} is the lifetime map of the donor in the presence of the acceptor, and τ_d is the average lifetime of the donor in the absence of the acceptor. The FRET efficiency maps indicate that, at localized donor lifetimes of 1.6 ns, the FRET efficiency increases to 30%. The efficiency scale is from 0% (dark blue) to 40% (dark red). (b) Quantification of localized EGFP lifetimes at the plasma membrane and the cytoplasm. An asterisk indicates significant reduction in EGFP lifetimes after EGF stimulation. A region of interest containing approximately 50 pixels was used to obtain a localized average lifetime value. The calculated EGFP lifetimes of the individual pixels were averaged (±SD) [Reproduced from Larijani (2003) with permission from *Curr. Biol.*]

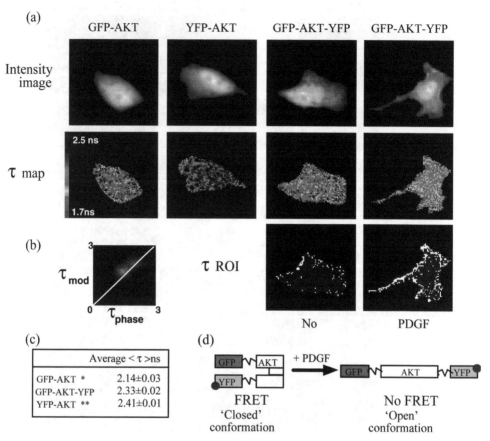

Figure 11.2   Change in GFP-AKT-YFP conformation upon cell stimulation. (a) Intensity images of NIH3T3 cells (top panels) and their corresponding average lifetime $\langle \tau \rangle$ maps (middle panels). NIH3T3 cells were transiently transfected with GFP–AKT, YFP–AKT and GFP–AKT–YFP fluorescent constructs. GFP–AKT–YFP transfected cells were treated for 5 min with 30 ng/ml of PDGF where indicated. The lifetime distribution is from 1.7 ± 0.15 ns to 2.5 ± 0.15. Upon stimulation there is a loss of FRET ('open' conformation) at the plasma membrane, indicated by the decrease in the GFP–AKT–YFP lifetime. The bottom panels represent two regions of interest (ROI) of lifetimes ranging from 1.7 to 2.0 ns (white pixels) and from 2.0 to 2.5 ns (blue pixels). (b) Two-dimensional histograms of (τ_p) and (t_m) lifetimes obtained in the frequency domain, illustrating the variation in lifetime. GFP–AKT (green dot), lowest lifetime; GFP–AKT–YFP (red dot), intermediate lifetime; and YFP–AKT (blue dot), highest lifetime ($n = 5$ cells). (c) Statistical analysis of the average lifetimes $\langle \tau \rangle$ of GFP–AKT, GFP–AKT–YFP and YFP–AKT. An asterisk indicates an extremely significant increase in the average lifetime of GFP–AKT–YFP compared with GFP–AKT with a p-value < 0.0003, $n = 8$ cells at 95% confidence interval. Two asterisks indicate a very significant difference between the average lifetime of GFP–AKTYFP vs YFP–AKT with a p-value < 0.0034, $n = 12$ cells at 95% confidence interval. (d) Schematic model of AKT conformational change upon PDGF stimulation. RET indicates a 'closed' conformation (with PDGF) and loss of RET an 'open' conformation (without PDGF). The red dot on YFP represents the red shifted version of EYFP (mutation Gln70Met)[Reproduced from Calleja *et al.*, 2003) with permission from *Biochem. J.*]

again at the molecular level such interactions can only be studied (noninvasively) by RET techniques.

11.6 Two-photon fluorescence

11.6.1 Basic concepts

Molecular two-photon absorption is the photophysical process whereby two photons with frequency ν are simultaneously absorbed by a molecule in a transition between an initial (ground) state and an (allowed) excited state that is raised from the ground state by energy $2h\nu$. The theoretical foundation of two-photon absorption was developed in 1931 by Maria Goppert-Mayer, although it was not until the development of short pulsed laser sources (nanosecond–picosecond) in the 1960s that two-photon absorption became a feasible experimental proposition. In simple terms, two-photon absorption can be thought of as two instantaneously consecutive single-photon transitions that take place via a set of 'virtual' intermediate states. Virtual states are those which can be coupled (allowed) by single photon transitions to the ground and excited states. Molecular two-photon absorption generally involves the absorption of red to near-infrared photons, wavelengths for which there are no single-photon electronic transitions. The virtual states are thus strongly off-resonance and as a result exist on a time scale Δt (lifetime) that is governed by the uncertainty principle

$$\Delta E \Delta t \approx \hbar \tag{11.1}$$

where

$$\Delta E = E_{\text{State}} - h\nu \tag{11.2}$$

The nearest electronic state that in principle could act as a virtual state would be the first excited (singlet) state with a transition energy corresponding to photons in the visible to ultraviolet regions. The energy gap ΔE is therefore on the order of $h\nu$; for a typical two-photon excitation wavelength of 800 nm these considerations yield a value of 4.2×10^{-16} s (0.42 fs). This is less than the oscillation period for light at 800 nm. Thus for two-photon absorption to occur, given of course a favourable quantum mechanical transition probability (see below), both photons must in effect interact with the molecule instantaneously, as illustrated in Figure 11.3. The probability of two photons interacting with a molecule simultaneously is dependent on the intensity of the light source employed together with the statistical nature of the light. This is quite distinct from single-photon transitions, which are not intensity-dependent, and as will be discussed later, this is the reason why two-photon-induced fluorescence offers significant advantages in biological applications such as fluorescence imaging.

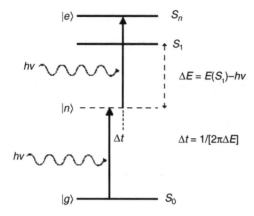

Figure 11.3 Schematic representation of molecular two-photon absorption between the molecular ground state $|g\rangle (S_0)$ and excited state $|e\rangle$ via a virtual state $|n\rangle$. The virtual state has a lifetime Δt determined by the uncertainty principle. For two-photon absorption to occur, the time difference between the two photons must be on or below Δt. The value of Δt is given by the difference between the photon energy and the nearest (single photon) resonant state which for an 800 nm photon is on the order of 0.42 fs

From quantum mechanical considerations the two-photon absorption rate (s^{-1}) is given by

$$R_{TPA} = \frac{\sigma^{(2)} I^2(t) G^{(2)}}{2(h\nu)^2} \qquad (11.3)$$

where $\sigma^{(2)}$ is the two photon absorption cross section (cm^4 s $photon^{-1}$), $I(t)$ is the instantaneous optical intensity (W cm^{-2}) and $G^{(2)}$ is the second order degree of coherence of the light. If $I(t)$ has a Gaussian temporal profile, integration of Equation (11.3) yields the transition probability

$$P_{TPA} = \frac{\sigma^{(2)}}{2(h\nu)^2} \left(\frac{E_{TP}}{A} \right)^2 \times \frac{0.664}{\tau} G^{(2)} \qquad (11.4)$$

where E_{TP} is the laser pulse energy (J), A is the area of illumination (cm^2) and τ is the full width (at half maximum) of $I(t)$ (s). The quantity $E_P/Ah\nu$ represents the photon flux (photons cm^{-2}) incident on the sample. From Equation (11.4) it can be seen that the probability of two-photon absorption is proportional to the square of the photon flux and inversely proportional to the width of the optical pulse. For a well mode-locked laser $G^{(2)} = 1$. Typical values for the molecular two-photon absorption cross section $\sigma^{(2)}$ range from 10^{-51} to 10^{-48} cm^4 s/photon (Table 11.1), although it will be seen later that it is possible to design and synthesize fluorescent chromophores with dramatically enhanced values of $\sigma^{(2)}$ over conventional fluorophores of comparable size.

Table 11.1 Two-photon cross sections of common fluorophores

Fluorophore	$\sigma^{(2)}$ ($\times 10^{-48}$cm^4 s/photon)	Excitation wavelength, λ (nm)
Extrinsic (synthetic) fluorophores		
Rhodamine 6G	1.16	700
Fluorescein	0.38	782
B-MSB	0.063	691
Indo-1 (free)	0.12	700
Intrinsic (naturally occurring) fluorophores		
Green Fluorescent Protein (GFP) wild type	0.06	800
NADH	0.0002	700
Serotonin	0.43	560
Dopamine	0.012	560

In the case of single photon absorption, the transition probability assuming excitation by an ultrashort laser pulse of energy E_{SP} is given by

$$P_{SPA} = \frac{\sigma E_{SP}}{h\nu_{SP}A} \qquad (11.5)$$

where σ is the single photon absorption cross-section (cm^2) and ν_{SP} is the transition frequency. For strongly allowed transitions typical values of σ range between 10^{-16} and 10^{-17} cm^2 (see Table 11.2). For single-photon absorption at 532 nm a transition probability of 10^{-3} with a cross section of 10^{-16} cm^2 requires a photon flux of 10^{13} photons cm^{-2}, corresponding to a pulse energy of 6.6×10^{-12} J and a spot size of 1.77×10^{-6} cm^2. For two-photon absorption, an equivalent transition probability, assuming a cross section of 2×10^{-48} cm^4 s/photon, requires, from Equation (11.4), a photon flux given by

$$\text{Flux} = \frac{E_{TP}}{Ah\nu} = \sqrt{\frac{10^{33}\tau_{(\text{picoseconds})}}{0.664}} \qquad (11.6)$$

A plot of Equation (11.6) is shown in Figure 11.4, where the enhancement of the two-photon transition probability using femtosecond laser pulses is evident. It is, however, worthwhile noting that a photon flux of 10^{16} cm^2 corresponds to a pulse energy of 4.33 nJ at 1.77×10^{-6} cm^2, which, using a 76 MHz mode-locked Ti:sapphire laser, would correspond to an on-sample power of 330 mW. Employing a more tightly focused beam, as for example, in the case of a fluorescence microscope, can reduce the need for high average on-sample excitation powers. Such arrangements allow significant two-photon absorption to be achieved with sub 100 mW laser powers. Nonetheless, it should be noted that the corresponding single photon flux is three orders of magnitude lower and many conventional molecular probes and biological fluorophores have cross sections lower than 10^{-48} cm^4 s/photon (see Table 11.1). In

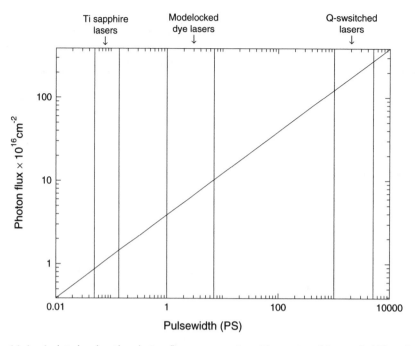

Figure 11.4   A plot showing the photon flux necessary to achieve a transition probability of 0.001 as a function of the excitation pulse width. Typical pulse width ranges for titanium sapphire, modelocked dye and Q-switched lasers are indicated. For 800 nm (a typical Ti sapphire wavelength) and taking a spot size of 1.77×10^{-6} cm^2, a photon flux of 10^{16} cm^{-2} corresponds to a pulse energy of 4.33 nJ

this light there has been considerable interest in developing molecular probes with significantly enhanced $\sigma^{(2)}$ values; these aspects of two-photon fluorescence will be discussed later.

11.6.2   Two-photon fluorescence

A schematic representation of molecular two-photon absorption and spontaneous emission is shown in the Jablonski diagram of Figure 11.5. Initial excitation from low-lying vibrational levels in the ground state (S_0) takes place via the simultaneous absorption of two (nonresonant) near-infrared photons, as discussed above. The initially excited state (S_n) undergoes rapid (sub-picosecond) internal conversion and vibrational relaxation in collisions with surrounding solvent molecules leading to a population of lower vibrational levels in S_1. In the absence of external perturbations, the population in S_1 decays by spontaneous emission (ca. 10^{-9} s) to upper vibrational levels of S_0 consistent with the Franck–Condon principle. Solvent deactivation of the vibrationally hot ground state population is correspondingly rapid. The fluorescence induced by molecular two-photon excitation originates from the same levels as are accessed by single photon excitation (Figure 11.6);

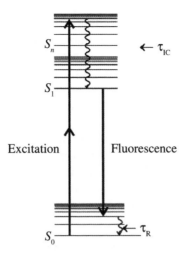

Figure 11.5 A Jablonski diagram indicating the excitation and relaxation pathways that lead to two-photon excited fluorescence. Excitation occurs from the molecular ground state S_0 to an excited singlet state S_n; this is followed by rapid (τ_{IC} sub-picosecond) internal conversion and collisional relaxation to lower vibrational levels of S_1. Spontaneous emission (nanoseconds) takes place to high vibrational levels of the ground state with fast (τ_R sub-picosecond) collisional relaxation returning molecules to the initially populated levels

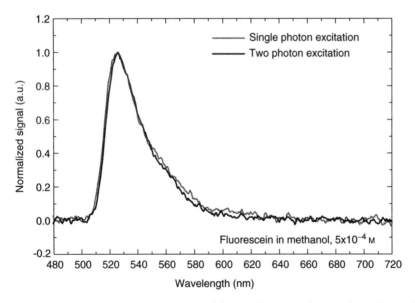

Figure 11.6 Steady-state fluorescence spectra of fluorescein (a standard single- and two-photon fluorophore) obtained using single-photon excitation at 500 nm and two-photon excitation at 800 nm. Although the two-photon transition energy is higher, 3.1 vs 2.48 eV, fast relaxation within the excited state manifold leads to the rapid population of the same low-lying vibrational levels in S_1 as evidenced by the equivalence of the emission spectra

here excitation is directly into high vibrational levels of S_1 with rapid solvent-mediated vibrational relaxation prior to fluorescence. In addition to an intensity-dependent transition probability, the excitation of molecular two-photon fluorescence differs from single photon excitation in that the absorption probability is polarization-dependent. A fundamental property of single-photon transitions in an isotropic medium is that the transition probability is independent of the polarization of the incident light; this restriction is relaxed in ordered samples where there is a direct correspondence between the molecular and laboratory frames of reference. However, for two-photon absorption the transition does not involve a simple dipolar rearrangement of electronic charge density but (in principle) involves products of the allowed single photon transition dipoles between the ground, intermediate and final states.

In quantum mechanical terms the cross-section for the absorption of two photons of polarisations α and β between states $|g\rangle$ and $|e\rangle$ is proportional to the squared modulus of the tensor $W^{g \rightarrow e}_{\alpha\beta}$ defined by a sum of transitions involving the intermediate (virtual) states $|n\rangle$,

$$|W^{g \rightarrow e}_{\alpha\beta}|^2 = \left| \sum_n \left(\frac{\alpha \cdot \langle g|r|n \rangle \langle n|r|e \rangle \cdot \beta}{\nu_n - \nu_\alpha + i\Gamma_n} + \frac{\beta \cdot \langle g|r|n \rangle \langle n|r|e \rangle \cdot \alpha}{\nu_n - \nu_\beta + i\Gamma_n} \right) \right|^2 \quad (11.7)$$

where ν_n and Γ_n are the transition frequency and homogeneous linewidth of $|n\rangle$ respectively. In the molecular frame of reference the transition tensor can have nine elements, S_{AB} where A, $B = X$, Y or Z,

$$|W^{g \rightarrow e}_{\alpha\beta}|^2 \equiv \left| \sum_{A,B=X,Y,Z} S_{AB} \right|^2 \quad (11.8)$$

For the absorption of two identical photons, $S_{AB} = S_{BA}$. For planar aromatic molecules the two-photon transition is dominated by moments in the plane of the molecule. The relative absorption strengths for linear and circular polarized two-photon absorption define the parameter Ω, which can be determined from the integrated fluorescence intensity for linear and circularly polarized absorption, as shown in Figure 11.4.

$$\Omega = I^C_{ABS}/I^L_{ABS} = \int I^C_F(t) \, dt \Big/ \int I^L_F(t) \, dt \quad (11.9)$$

The total fluorescence intensity can be obtained from a 'magic angle' polarization measurement at 54.7° to the symmetry axis (axis of cylindrical symmetry for the transition). In terms of a planar transition tensor where the emission transition dipole moment lies along the molecular axis, Ω is given by

$$\Omega = \frac{1}{2} \frac{(S_{XX} + S_{YY})^2 + 3(S_{XX} - S_{YY})^2 + 12S_{XY}^2}{2(S_{XX} + S_{YY})^2 + (S_{XX} - S_{YY})^2 + 3S_{XY}^2} \tag{11.10}$$

For a diagonal transition tensor $S_{XY} = 0$, Ω and $R(0)$ are determined solely by the ratio $S = S_{YY}/S_{XX}$,

$$\Omega = \left(\frac{1}{2}\right) \frac{(1+S)^2 + 3(1-S)^2}{2(1+S)^2 + (1-S)^2} \tag{11.11}$$

Values for Ω can vary from $3/2(S=-1)$ to $1/4$ $(S=1)$; a value of $\frac{2}{3}$ corresponds to a diagonal transition tensor for which $S = 0$. The polarization of two-photon fluorescence is an important experimental quantity. Measurement of the initial fluorescence anisotropy – the 'time–zero' degree of polarization of the spontaneous emission yields information on the degree of order in aligned systems and fast depolarization mechanisms such as energy transfer and fast orientational relaxation. Fluorescence anisotropy measurements also yield information on the local structure of chromophores in proteins and the structure of biological macromolecules. In this light it is worth noting that the initial fluorescence anisotropy is also dependent on the structure of the transition tensor, for linearly polarized excitation the initial fluorescence anisotropy as measured in Figure 11.7 is given by

$$R(0) = \frac{I_V(0) - I_H(0)}{I_V(0) + 2I_H(0)} = \left(\frac{1}{7}\right) \frac{2(S_{XX} + S_{YY})^2 + (S_{XX} - S_{YY})^2 + 4S_{XY}^2 + 9(S_{XX}^2 - S_{YY}^2)}{2(S_{XX} + S_{YY})^2 + (S_{XX} - S_{YY})^2 + 3S_{XY}^2} \tag{11.12}$$

For a diagonal transition tensor

$$R(0) = \left(\frac{1}{7}\right) \frac{2(1+S)^2 + (1-S)^2 + 9(1-S^2)}{2(1+S)^2 + (1-S)^2} \tag{11.13}$$

The variation in Ω and $R(0)$ with S is displayed in Figure 11.8. Where the transition is largely dominated by S_{XX}, an initial fluorescence anisotropy of $\frac{4}{7}$ is obtained that is characteristic of a $\cos^4\theta$ distribution of transition dipoles in the laboratory frame; the corresponding value for Ω is $\frac{2}{3}$. Deviations from these values are indicative of contributions from S_{YY} and/or S_{XY}.

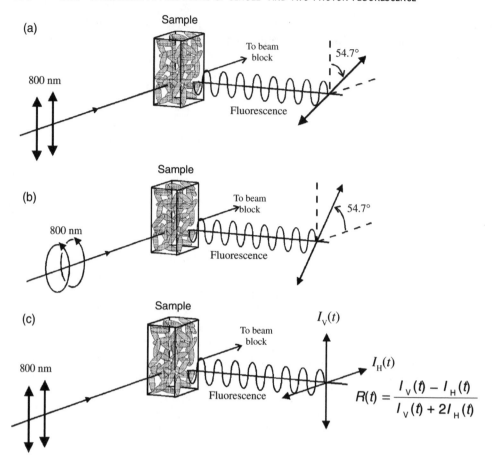

Figure 11.7   Excitation-detection geometries for the determination of the transition strength for (a) linearly polarized and (b) circularly polarized two-photon excitation. Fluorescence intensity measurements are made at magic angle polarization settings with respect to the quantization (symmetry) axis for the two polarizations. Alternatively, the fluorescence intensity can be constructed from $I_V V(t) + 2I_H(t)$ and $2I_V(t) + I_H(t)$ measurements for linear and circular polarizations respectively. (c) Excitation–detection geometry for fluorescence anisotropy measurements following linearly polarized two-photon excitation

11.7   Applications of two-photon fluorescence

11.7.1   Two-photon fluorescence imaging

Perhaps the most significant contribution that two-photon excited fluorescence has made in biological research to date is in the area of fluorescent imaging. Prior to the invention of two-photon microscopy in 1990, depth resolution in fluorescence microscopy was achieved using the confocal microscope, the principles of which are illustrated in Figure 11.9. Fluorescence is excited via single-photon excitation

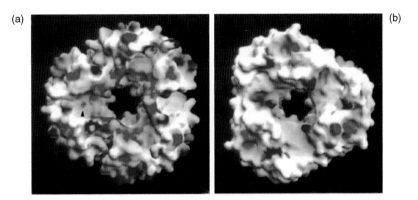

Figure 2.5

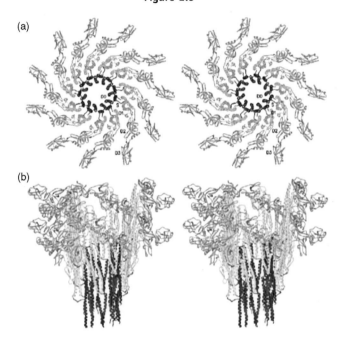

Figure 2.6

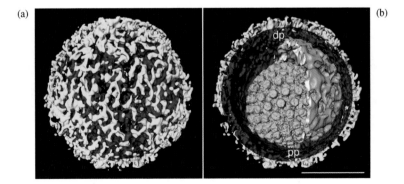

Figure 2.12

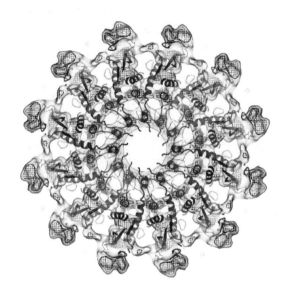

Figure 2.8

(a)   (b)

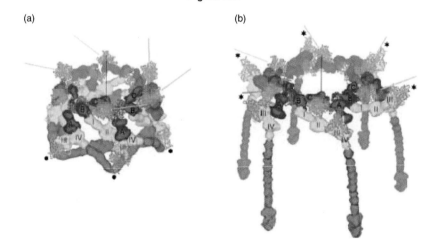

Figure 2.9

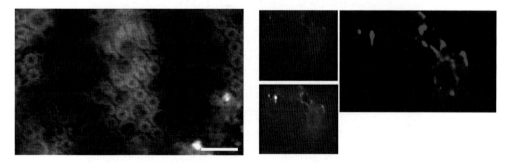

Figure 3.4   Figure 5.7

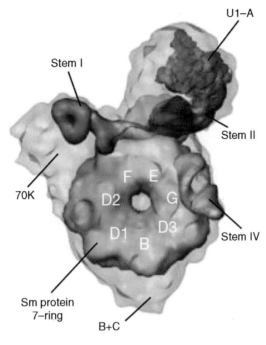

Figure 2.10

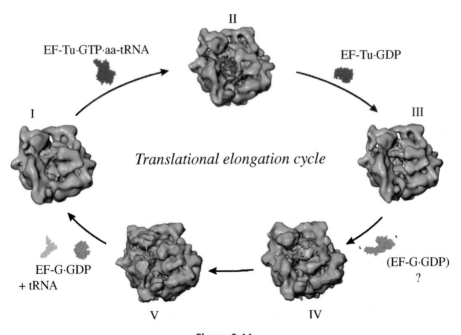

Translational elongation cycle

Figure 2.11

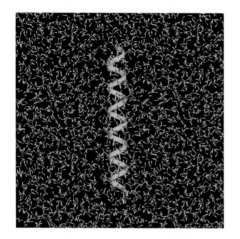

Figure 9.5

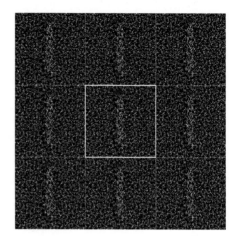

Figure 9.6

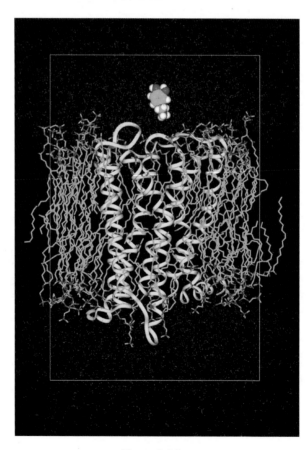

Figure 9.16

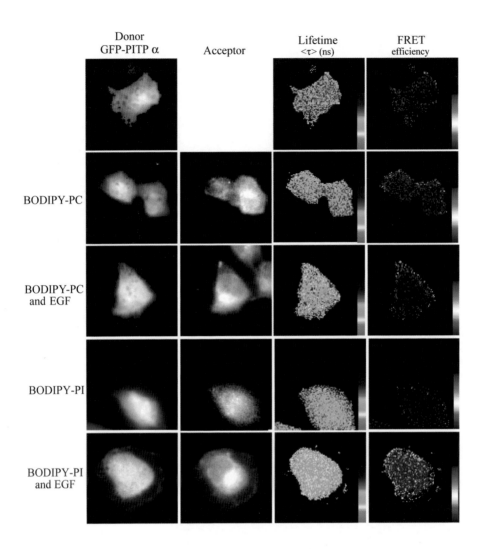

	Donor GFP-PITP α	Acceptor	Lifetime ⟨τ⟩ (ns)	FRET efficiency
BODIPY-PC				
BODIPY-PC and EGF				
BODIPY-PI				
BODIPY-PI and EGF				

	GFP lifetime (τ ns)	
	Cytoplasm	Plasma Membrane
GFP-PITPα	2.02±0.16	2.03±0.15
GFP-PITPα +BODIPY-PC	2.16±0.12	2.11±0.07
GFP-PITPα +BODIPY-PC +EGF	2.00±0.15	1.75±0.13*
GFP-PITPα +BODIPY-PI	2.16±0.12	1.90±0.10
GFP-PITPα + BODIPY-PI +EGF	1.82±0.09*	1.60± 0.10*

Figure 11.1

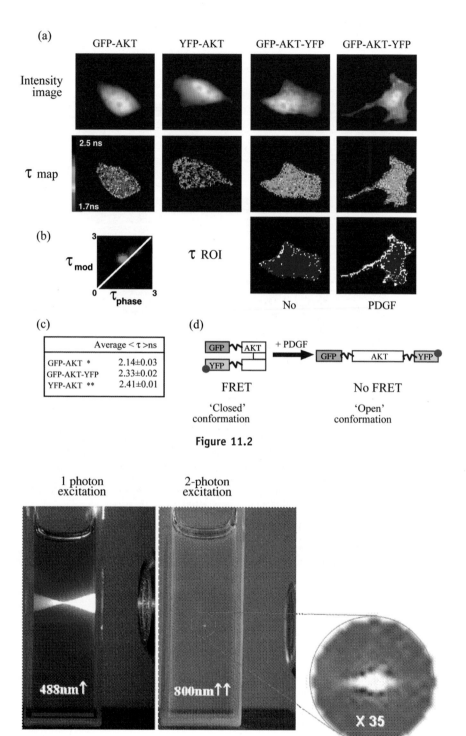

(a)

| GFP-AKT | YFP-AKT | GFP-AKT-YFP | GFP-AKT-YFP |

Intensity image

τ map

2.5 ns

1.7ns

(b)

τ ROI

3

τ_mod

0   τ_phase   3

No   PDGF

(c)

	Average < τ >ns
GFP-AKT *	2.14±0.03
GFP-AKT-YFP	2.33±0.02
YFP-AKT **	2.41±0.01

(d)

GFP — AKT
YFP —

+ PDGF →

GFP — AKT — YFP

FRET   No FRET

'Closed'   'Open'
conformation   conformation

Figure 11.2

1 photon
excitation

2-photon
excitation

488nm↑

800nm↑↑

X 35

Figure 11.11

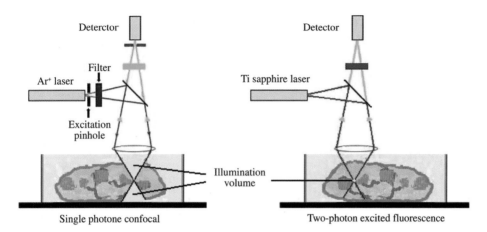

Single photone confocal

Two-photon excited fluorescence

Figure 11.12

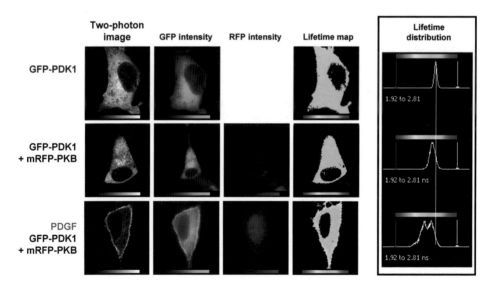

Figure 11.15

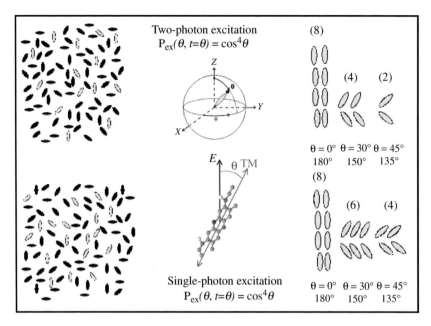

Two-photon excitation
$$P_{ex}(\theta, t=0) = \cos^4\theta$$

(8)

(4) (2)

$\theta = 0°$ $\theta = 30°$ $\theta = 45°$
$180°$ $150°$ $135°$

(8)

(6) (4)

Single-photon excitation
$$P_{ex}(\theta, t=0) = \cos^4\theta$$

$\theta = 0°$ $\theta = 30°$ $\theta = 45°$
$180°$ $150°$ $135°$

Figure 11.16

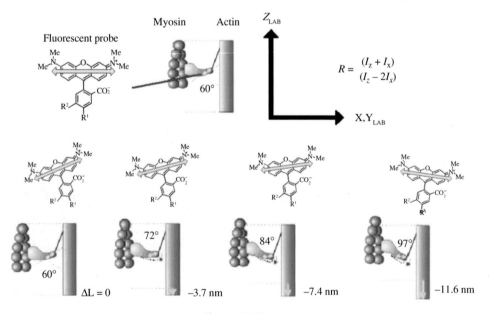

Fluorescent probe   Myosin   Actin Z_{LAB}

$$R = \frac{(I_z + I_x)}{(I_z - 2I_x)}$$

$60°$

X, Y_{LAB}

$60°$ $72°$ $84°$ $97°$

$\Delta L = 0$ -3.7 nm -7.4 nm -11.6 nm

Figure 11.19

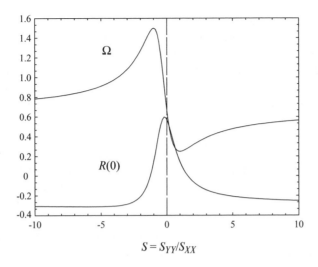

Figure 11.8 Variation in the absorption ratio Ω and the initial anisotropy $R(0)$ with the ratio $S = S_{YY}/S_{XX}$ for a diagonal planar two-photon transition tensor. The dashed line at $S = 0$ intersects the curves at $\Omega = 2/3$ and $R(0) = 4/7$

using a continuous wave laser (usually one of the lines of an argon ion laser) which is scanned in the *X–Y* plane to build up an image of the sample. However as considerable fluorescence is excited both above and below the focal plane it will, if allowed to reach the detector, cause blurring of the final image. In confocal

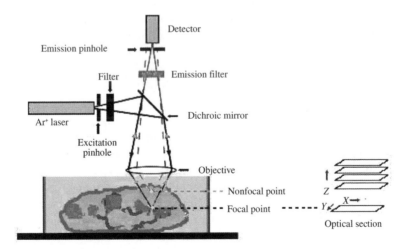

Figure 11.9 Schematic illustration of the principles of confocal fluorescence microscopy; single-photon excited fluorescence occurs along the entire beam path, however the insertion of a pinhole in the focal plane of the objective ensures that fluorescence from outside the focal point of the laser (dashed line) is prevented from reaching the detector. The laser is scanned in the *X–Y* plane to build up an image of the sample for a given depth. Translation of the sample in the *Z*-direction allows images (optical sections) to be recorded as a function of depth

microscopy, a pinhole is introduced in front of the detector at a position that excludes emission from both above and below the focal plane of the sample; this condition is satisfied when the focal plane of illumination and detection are the same. The scanning confocal microscope is widely used to obtain sharp images or optical sections of a specimen by scanning laser in the horizontal plane for a given sample depth. The drawbacks of such an instrument are that large regions of the sample are excited; this carries a significant risk of damage (particularly for studies of live cells) arising from heat deposition and photochemical reactions. In addition the introduction of the confocal pinhole results in reduced collection efficiency from the region of interest with a subsequent increase in data collection time. As will be seen, two-photon excitation is ideally suited for scanning fluorescence microscopy, allowing confocal resolution with a significant reduction in the drawbacks outlined above.

From Equation (11.3) it can be seen that the two photon absorption probability is dependent on the incident laser intensity. Thus two-photon absorption will take place most readily at the laser focus. The transverse (spatial) dependence of a the output intensity of a laser operating in a fundamental (lowest loss) mode at time t has the form

$$I(z, r, t) = \frac{2P(t)}{\pi [w(z)]^2} \exp\left(-\frac{2r^2}{[w(z)]^2}\right) \tag{11.14}$$

where z is the distance along the optical axis, r is the radial distance away from the optical axis, $P(t)$ is the incident power and $w(z)$ is a quantity known as the spot size of the beam. The spot size at any given point (z) determines the radial intensity dependence of the laser. From a minimum value (w_0), determined by internal and external cavity optics, the spot size increases according to

$$w(z) = w_0 \sqrt{1 + (z/z_R)^2} \tag{11.15}$$

where the quantity z_R, known as the Rayleigh range, determines the focal depth of the Gaussian beam; the Rayleigh range is the distance over which $w(z) \leq w_0$ and physically represents the region where the beam remains effectively collimated. The two-photon transition probability, from Equation (11.3), is inversely proportional to the square of the instantaneous intensity and hence proportional to $w(z)^{-4}$. A plot of $w(z)$ and the corresponding two-photon transition probability is shown in Figure 11.10, approximately 82 per cent of the two-photon absorption takes place within the Rayleigh range, leading to an effective focal volume of 32×10^{-15} l (32 fl). The consequence of this is a dramatic increase in axial resolution as compared with single-photon excited fluorescence where excitation occurs throughout the illuminated sample volume, as can be seen in Figure 11.11. In addition to increased spatial resolution in three-dimensional imaging, the restriction of two-photon excitation to femtolitre volumes leads to a greatly reduced risk of sample photo-degradation. The use of near-infrared excitation in two-photon excitation, as opposed to visible and

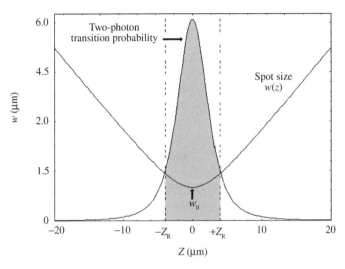

Figure 11.10 A plot of $w(z)$ and the corresponding spatial dependence of the two-photon transition probability for an excitation wavelength of 800 nm and a minimum spot size of 1 μm. Approximately 82% of the two-photon absorption takes place within the Rayleigh range $\pm Z_R$ about the focal point at $Z = 0$, leading to an effective focal volume of ca. 32×10^{-15} litres (32 fl)

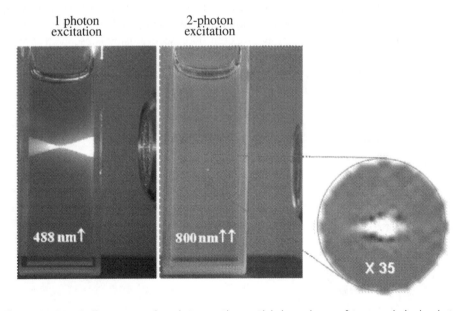

Figure 11.11 A direct comparison between the spatial dependence of two- and single-photon absorption from a fluorescent probe in an isotropic solution contained in a cuvette and viewed at 90° to the excitation direction. Single-photon excitation at 488 nm produces fluorescence throughout the sample, whereas the restriction of two-photon fluorescence (excited at 800 nm) to the Rayleigh range around the laser focus is evident. The photographs were provided courtesy of Dr Mireille Blanchard-Desce (CNRS UMR 6510 Rennes)

ultraviolet wavelengths, in single-photon excitation is in itself less detrimental to studies of living cells. In tissues better penetration is afforded due to lower scattering losses and (single-photon) absorption by endogenous molecules; the 'biological window' between 700 and 1200 nm coincides with the two-photon absorption bands of many fluorescent probes (Table 11.1). A two-photon scanning microscope thus provides near-confocal resolution with increased fluorescence collection efficiency and a reduction in sample damage; a comparison between the scanning confocal microscopy and two photon microscopy techniques is illustrated in Figure 11.12. The basic components of a two-photon scanning microscope system are set out in Figure 11.13. As well as providing a step change in the biologist's ability to noninvasively probe the structure of biological specimens with submicron resolution, two-photon excitation also finds application as a means of investigating localized dynamical processes. At the cellular level solutes are moved by diffusion and most conventional approaches for measuring diffusion such as radioisotope tracers have insufficient spatial or temporal resolution to measure diffusion coefficients with submicron resolution. In recent years optical techniques, such as fluorescence recovery after photo-bleaching (FRAP) and wide field flash photolysis have been shown to yield increased spatial resolution. However problems arise from out-of-focus fluorescence events which can contaminate the in-plane recordings of fluorescence change. Two-photon excitation provides a three-dimensionally resolved source of excitation which, in conjunction with a suitable photo-activatable compound, can provide a microscopic 'point source' with which to probe local diffusive properties. Photo-activation is achieved by un-caging otherwise inert molecules whose spatial migration from the laser focal volume is monitored by two-photon fluorescence using subsequent 'probe' pulses; the principles of the technique are illustrated in Figure 11.14. In addition to studies of diffusion dynamics, photochemical un-caging has been used to study calcium ion transport and has been applied to studies of intracell signalling.

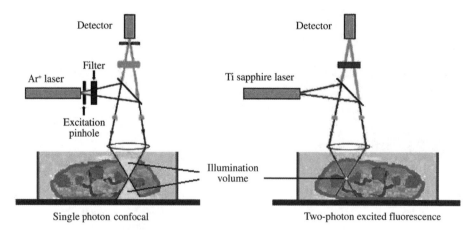

Figure 11.12 A comparison between confocal and two-photon fluorescence microscopy; in confocal microscopy fluorescence is excited throughout the sample while in two-photon excitation it is restricted to the confocal region around the focal point of the laser

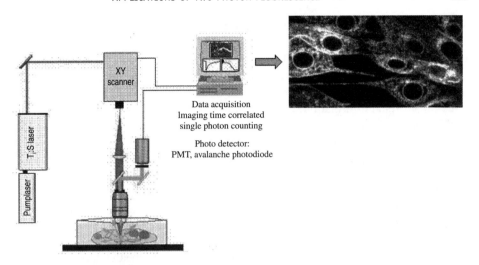

Figure 11.13 Schematic representation of a two-photon fluorescence microscope system; two-photon excitation of fluorophores within the sample is achieved using the output of a Ti:sapphire laser. This is scanned in the X–Y plane to produce an image of a section of the sample. The insert shows a two-photon fluorescence image of pig kidney cells labelled with two-photon fluorophores (image provided courtesy of Dr Mireille Blanchard-Desce (CNRS UMR 6510 Rennes))

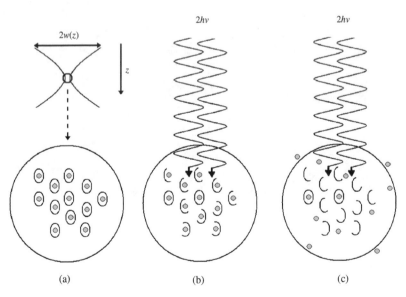

Figure 11.14 Photochemical un-caging using two-photon excitation. Inactive ('caged') fluorescent probes within the focal volume of the laser (a) are first activated by two-photon-induced photochemical un-caging (b); diffusion of the probes from the focal volume is then monitored by subsequent two-photon-induced fluorescence

11.7.2   Biological example – two-photon time-domain FLIM

Cells transfected with GFP fused to the protein of interest can be excited by two-photon absorption of near-infrared light. As mentioned in the previous section this method of excitation reduces photo-damage and acquires inherently z-sectioned, three-dimensional information. This is achieved without the use of confocal apertures as this nonlinear absorption process is confined to a single, diffraction-limited focal plane. Photodamage is thus minimized, permitting long periods of data acquisition.

The next example illustrates how two-photon time domain FLIM can be used for detecting the interaction of two protein kinases PDK1 (3,4,5-phosphoinositide protein kinase) and PKB (protein kinase B) at the plasma membrane of NIH3T3 cells. Both PDK1 and PKB associate with PtdIns(3,4,5)P$_3$ and PtdIns(3,4)P$_2$ via their plecktrin homology (PH) domains. It seems that this mutual interaction with such lipids leads these enzymes to co-localize at the plasma membrane and in turn to activate PKB. However, until recently it had not been shown that these molecules actually associate at the plasma membrane.

The exploitation of RET by two-photon time domain FLIM has permitted illustration of the association of these two kinases. Figure 11.15 shows that NIH3T3 is

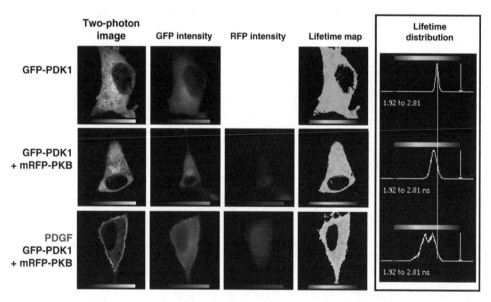

Figure 11.15   Interaction of PDK1 and PKB detected by two-photon time domain FLIM. NIH3T3 are transfected with GFP-PDK1 (upper panel), co-transfected with mRFP-PKB (middle panel) and stimulated by growth factor PDGF (lower panel). The lifetime maps indicate that the GFP-PDK1 lifetime changes at the plasma membrane of these cells upon stimulation. In the presence of the acceptor mRFP-PKB there is no variation of the donor lifetime (GFP-PDK1) at the plasma membrane. The lifetime distributions are indicated by the histograms (right panels). It can be clearly seen that, upon stimulation, the GFP-PDK1 lifetime at the plasma membrane decreases from 2.5 to 1.9 ns and the GFP-PDK1 lifetime at the cytoplasm (2.3 ns) remains the same as when the acceptor is present. The decrease in lifetime at the plasma membrane illustrates that PDK1 and PKB associate upon growth factor stimulation

transfected by GFP-PDK1 alone (donor chromophore) and/or with mRFP-PKB (acceptor chromophore), as indicated. Upon stimulation with a growth factor (PDGF), both proteins translocate to the plasma membrane. In the presence of the acceptor there is a clear shift of the donor lifetime from 2.4 to 1.9 ns. This decrease in lifetime indicates that RET has occurred. In order for RET to occur, the donor–acceptor pairs need to be at a distance inferior to 7 nm. The change in the donor lifetime does not occur at the plasma membrane without the acceptor or without PDGF stimulation. Note that, with two-photon, unlike single-photon frequency-domain FLIM, the plasma membrane is precisely observed.

11.8   Photoselection and fluorescence anisotropy

Time-resolved molecular fluorescence finds wide application in the biological sciences as a probe of molecular structure, environment and dynamics. Fluorescent molecular probes are widely used to study the properties of host environments ranging from ordered molecular materials (e.g. liquid crystals) to surfaces, cells, macromolecular assemblies (e.g. proteins) and pseudo-domain structures in complex fluids. Fluorescence lifetime modification (i.e. shortening) arising from Förster resonance energy transfer (RET) between excited donor and acceptor molecules is an important quantitative indicator of short-range molecular order and conformational change. Two-photon fluorescence measurements provide additional information to that provided by single-photon excitation, first an increased degree of orientational photo-selection, that is, a more precise angular selection of fluorescent molecular orientations and, second, information that single photon measurements cannot provide.

11.9   Fluorescence anisotropy and isotropic rotational diffusion

As discussed above, for a single element planar transition tensor the angle-dependent two-photon transition probability is the squared modulus of the single photon $\cos^2 \theta$ transition probability. Spontaneous emission from a two-photon excited molecular population occurs via the same $S_1 \rightarrow S_0$ transition as in single photon fluorescence; however, the evolving orientational distribution of molecules reflects the increased degree of order created in the excited state by the excitation process.

The higher initial anisotropy reflecting the greater proportion of molecules selected that are aligned parallel to the excitation polarization in two-photon excitation is illustrated schematically in Figure 11.16. The evolution of the fluorescence anisotropy is influenced by fast processes such as energy transfer; however for isolated (noninteracting) molecules in isotropic media, the predominant mechanism is a loss of order due to oreintational averaging in numerous (ca. 10^{12}–10^{13} s^{-1}) small steps due to collisions with surrounding solvent molecules. If the emission and absorption transition dipole moments are parallel and lie along (or close to) a

symmetry axis of the molecule, then the time-dependence of the fluorescence anisotropy can be described in terms of single axis rotational diffusion,

The components of the fluorescence anisotropy measured as in Figure 11.7c
following excitation with a short [compared with fluorescence emission and
orientational relaxation (molecular motion) in the excited molecules] laser
pulse are given by

$$I_Z(t) = N_{ex}(t)[1 + 2R(t)] \tag{11.16}$$
$$I_Y(t) = N_{ex}(t)[1 - R(t)] \tag{11.17}$$

where $N_{ex}(t)$ is the excited state population remaining at time t following
excitation and $R(t)$ is the fluorescence anisotropy of the excited state population. $R(t)$ represents the degree of order remaining in the excited state following
the excitation process. From simple symmetry considerations it can be shown
that $R(t)$ is proportional to the degree of second-order alignment present in the
excited state

$$R(t) \equiv \langle \alpha_{20}^{ex}(t) \rangle / \sqrt{5} = \sqrt{\frac{4\pi}{5}} \int_0^{2\pi} \int_0^\pi Y_{20}^*(\theta, \phi) P_{ex}(\theta, \phi, t) \sin\theta \, d\theta \, d\phi \tag{11.18}$$

Here $P_{ex}(\theta, \phi, t)$ is the probability at time t of finding the a molecule aligned
between the polar angles θ to $\theta + d\theta$ and ϕ to $\phi + d\phi$ (Figure 11.16) and
$Y_{20}(\theta, \phi)$ is a spherical harmonic whose form is

$$Y_{20}^*(\theta, \phi) = \sqrt{\frac{5}{16\pi}}(3\cos^2\theta - 1) \tag{11.19}$$

Inserting Equation (11.18) into Equation (11.19) yields

$$R(t) \equiv \frac{1}{2}[3\langle\cos^2\theta(t)\rangle - 1] \tag{11.20}$$

where $\langle\rangle$ denotes an average over the distribution. For an isotropic distribution $\langle\cos^2\theta\rangle = \frac{1}{3}$, and the fluorescence anisotropy is necessarily zero; for a
$\cos^2\theta$ distribution this value is $\frac{3}{5}$ corresponding to an initial single-photon
excited fluorescence anisotropy of $\frac{2}{5}$; in the case of two-photon excitation with
$S = 0$ (Equation 11.13) the corresponding values are $\frac{5}{7}$ and $\frac{4}{7}$ respectively.

$$\frac{\partial P_{ex}(\theta, \phi, t)}{\partial t} = D\left(\frac{\partial^2}{\partial x^2} + \frac{\partial^2}{\partial y^2} + \frac{\partial^2}{\partial z^2}\right)P_{ex}(\theta, \phi, t) \tag{11.21}$$

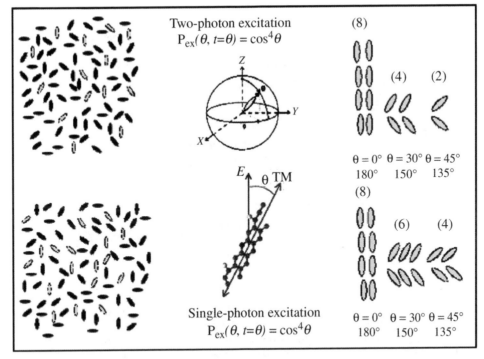

Figure 11.16 A schematic representation of single- and two-photon photoselection of fluorescent probe molecules in an isotropic medium. For a single element transition tensor lying along the symmetry axis of the fluorescent probe the angular dependence of the transition probability in the laboratory frame is proportional to $\cos^4 \theta$, where θ is the angle between the transition moment and the polarization of the laser E. For a similarly oriented single photon transition dipole, the transition probability shows a $\cos^2 \theta$ dependence. Molecules that make a small angle with respect to $E(Z)$ are preferentially photoexcited over those that are oriented more closely to the X–Y plane. In two-photon excitation this angular dependence is more pronounced, leading to a lower population of excited molecules in the region between $\theta = 0°$ and $180°$

where D is the rotational diffusion coefficient (s^{-1}). For the $\langle \alpha_{20}^{ex}(t) \rangle$ moments of the excited state distribution this yields

$$\langle \alpha_{20}^{ex}(t) \rangle = \langle \alpha_{20}^{ex}(t = 0) \rangle \exp(-6Dt) \tag{11.22}$$

The fluorescence anisotropy (from 11.18) is thus

$$R(t) = R(t = 0) \exp(-t/\tau_{ROT}) \tag{11.23}$$

where τ_{ROT} is the rotational diffusion time given by

$$\tau_{ROT} = 1/6D \tag{11.24}$$

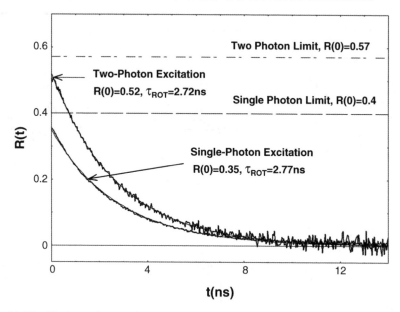

Figure 11.17   Single and two photon excited fluorescence anisotropy decays recorded for rhodamine 6G in ethylene glycol, the initial anisotropies for both processes are close to the theoretical maxima (dashed lines) for excitation from an isotropic ground state and (in the case of the two-photon excited population) a diagonal transition tensor dominated by S_{xx}

The rotational diffusion time is proportional to the molecular hydrodynamic volume V and solvent viscosity η and inversely proportional to the thermal energy kT

$$\tau_{\text{ROT}} = fC(\eta V/kT) \qquad (11.25)$$

This relationship is modified by two constants: the molecular shape factor f (a function of the molecular dimensions) and the boundary coefficient C, which takes into account the interaction between the solvent and the solute. In principle, two-photon fluorescence anisotropy decays in isotropic media should yield the same diffusion times as for single photon excitation, but with significantly increased initial fluorescence anisotropy; this can be seen in Figure 11.17, which compares single- and two-photon anisotropy decays for the fluorescent probe rhodamine 6G in ethylene glycol. Rotational diffusion times for small molecular probes vary from nanoseconds to hundreds of picoseconds for isotropic rotational diffusion in low viscosity solvents.

11.10   Fluorescent probes in proteins and membranes

The rotational dynamics of naturally occurring fluorescent probes found in large molecules such as proteins or that are used to probe structured environments such as

membranes are more complex than the unrestricted rotational diffusion discussed above. To begin with, the fluorescent probe is no longer 'free' but is
confined to some restricted geometry due to the local environment. The steady-
state *local* orientational distribution of the fluorophore is no longer isotropic;
however the global orientational distribution of the larger host molecule is wholly
random and the fluorescence anisotropy will tend towards zero. The two orientation relaxation processes can be separated by several orders of magnitude; for
tryptophan 214 in human serum albumin (HSA) the local tryptophan motion
has a rotational lifetime on the order of 200 ps whilst HSA shows a rotational
diffusion time of 20–30 ns depending on the biological molecules it transports. If
the two orientational relaxation processes are independent, the fluorescence
anisotropy is described by

$$R(t) = \exp\left(\frac{-t}{\tau_{\text{SLOW}}}\right)\left[A\exp\left(\frac{-t}{\tau_{\text{FAST}}}\right) + B\right] \tag{11.26}$$

The pre-exponential factors in Equation (11.26) can be used to provide information
on the local order of the fluorescent probe which is modelled in terms of restricted
diffusion within a cone of semi-angle α, as shown in Figure 11.18.

$$\sqrt{\frac{A - B}{A + B}} = \frac{1}{2}[\cos\alpha(1 + \cos\alpha)] \tag{11.27}$$

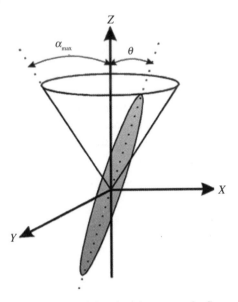

Figure 11.18 A representation of the restricted environment of a fluorescent probe embedded in
a larger species. The fluorescence anisotropy of the probe exhibits a fast decay due to angular
averaging within a cone of angular width α_{MAX}. This decay is mediated by the slower overall
rotational diffusion of the host

Two photon fluorescence anisotropy measurements yield anisotropy decays of a wholly equivalent form to Equation (11.26) but with larger values of A and B and a consequent reduction in experimental uncertainty for α. A recent application of the technique has been the investigation of local domain structure in the isotropic phase of liquid crystalline materials. Incorporation of a fluorescent chromophore within a large biological molecule does not always lead to a biexponential anisotropy decay, in the case of GFP the rigid cage imposed by the protein both stabilizes the fluorophore and prevents any significant local motion. Two-photon fluorescence anisotropy measurements of wild-type GFP show an approximately single exponential decay of 16 ± 1 ns corresponding to protein orientation relaxation and an initial anisotropy close to $\frac{4}{7}$.

11.10.1   Ordered molecular systems

Considerable information on molecular interactions and molecular structure can be obtained from studies of ordered materials and samples; here the degree of averaging between laboratory and molecular frames of reference is either wholly removed or greatly reduced. There is now a net degree of molecular order in the laboratory frame of reference. As in the case of probe fluorescence [see Equation (11.18)], this *intrinsic* distribution is conveniently expressed as an expansion in spherical harmonics,

$$P_{\text{LAB}}(\theta, \phi) = \sum_{KQ} \langle C_{KQ} \rangle Y_{KQ}(\theta, \phi) \tag{11.28}$$

For a cylindrically symmetric molecular array (no net ordering in the X–Y plane) $Q = 0$, under these circumstances expanding Equation (11.26) and normalizing to the scalar $Y_{00}(\theta, \phi)$ term, the first four terms of the distribution function have the form

$$P_{\text{LAB}}(\theta, \phi) = \frac{1}{4\pi} \left[1 + \begin{array}{l} \left[\langle \alpha_{20} \rangle \frac{\sqrt{5}}{2} (3\cos^2 \theta - 1) + \langle \alpha_{20} \rangle \frac{3}{8} (35 \cos^4 \theta - 30 \cos^2 \theta + 3) \right. \\ \left. + \langle \alpha_{60} \rangle \frac{\sqrt{13}}{16} (231 \cos^6 \theta - 315 \cos^4 \theta + 105 \cos^2 \theta - 105) + \ldots \right] \end{array} \right] \tag{11.29}$$

where $\langle \alpha_{KQ} \rangle = \langle C_{KQ} \rangle / \langle C_{00} \rangle$.

In an analogous manner to single-photon absorption it is possible to detect molecular order (the $\langle \alpha_{KQ} \rangle$ parameters) by the difference in the absorption of Z and X (or Y) polarized light. The absorption of Z and X

polarized light in single-photon transitions the components of $P_{LAB}(\theta, \phi)$ differently:

$$I_Z = cN_{PROBE} \int\limits_0^{2\pi} \int\limits_0^{\pi} P_{LAB}(\theta, \phi) \cos^2\theta \sin\theta \, d\theta \, d\phi \qquad (11.30)$$

$$I_X = cN_{PROBE} \int\limits_0^{2\pi} \int\limits_0^{\pi} P_{LAB}(\theta, \phi) \sin^2\theta \cos^2\phi \sin\theta \, d\theta \, d\phi \qquad (11.31)$$

where N_{PROBE} is the number of probe molecules and c is a constant of proportionality. Measurement of the ratio $(I_Z - I_X)/(I_Z + 2I_X)$ yields

$$\frac{I_Z - I_X}{I_Z + 2I_X} = \frac{\langle\alpha_{20}\rangle}{\sqrt{5}} \qquad (11.32)$$

This approach has been widely used to study biological systems, notably to detect tilting motions of myosin domains in muscle fibres. Measurement of the $\langle\alpha_{20}\rangle$ order parameter alone does not necessarily yield a complete picture of molecular order, and neglect of higher terms in Equation (11.28) can lead to serious errors. Two-photon dichroism measurements are, however, sensitive to $\langle\alpha_{20}\rangle$ and the next term in the expansion $\langle\alpha_{40}\rangle$. The absorption strengths for Z and X polarized two-photon absorption are given by

$$I_Z = dN_{PROBE} \int\limits_0^{2\pi} \int\limits_0^{\pi} P_{LAB}(\theta, \phi) \cos^4\theta \sin\theta \, d\theta \, d\phi \qquad (11.33)$$

$$I_X = dN_{PROBE} \int\limits_0^{2\pi} \int\limits_0^{\pi} P_{LAB}(\theta, \phi) \sin^4\theta \cos^4\phi \sin\theta \, d\theta \, d\phi \qquad (11.34)$$

where d is a constant of proportionality. The corresponding absorption ratio is given by

$$\frac{I_Z - I_X}{I_Z + 2I_X} = \frac{\frac{10\langle\alpha_{20}\rangle}{7\sqrt{5}} + \frac{5}{63}\langle\alpha_{40}\rangle}{1 + \frac{2}{9}\langle\alpha_{40}\rangle} \qquad (11.35)$$

Two photon dichroism measurements have been used to study the structural properties of highly ordered liquid crystal mesophases; from measurements of $\langle\alpha_{20}\rangle$ and $\langle\alpha_{40}\rangle$ it is possible [given a realistic model for $P_{LAB}(\theta, \phi)$] to predict values of higher terms. Knowledge of $\langle\alpha_{20}\rangle$ and

$\langle\alpha_{40}\rangle$, however, represents the necessary minimum information to make an accurate prediction of molecular order. Determination of $\langle\alpha_{20}\rangle$ and $\langle\alpha_{40}\rangle$ is possible in fluorescence anisotropy experiments, time-resolved single-photon fluorescence anisotropy measurements allow the measurement of $\langle\alpha_{20}\rangle$ and $\langle\alpha_{40}\rangle$ through the initial fluorescence anisotropy. It has been recently shown that measurements of $\langle\alpha_{60}\rangle$ are possible from two-photon fluorescence anisotropy measurements, allowing a more accurate determination of molecular order and also as a critical test of models for $P_{LAB}(\theta, \phi)$. Polarized fluorescence measurements have also been widely applied to studies of molecular motion in skeletal muscle fibres; aligned samples are tagged with fluorescent probes and anisotropy changes are measured in a polarizing microscope, yielding angular information on the probe order during extension and contraction and hence the relative motion of the components of the system, as illustrated in Figure 11.19.

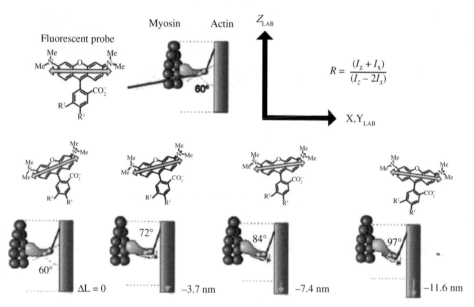

Figure 11.19 A schematic illustration of the application of fluorescence anisotropy measurements to study orientation changes in skeletal muscle fibres. A rhodamine-based fluorescent probe is bound to the regulatory light chain (RLC) of the myosin head. Changes in probe orientation directly related to changes in the steady-state orientation of the emission transition dipoles (yellow arrows) of the fluorescent probes in the laboratory frame which are monitored by fluorescence anisotropy. Stretches of the muscle fibre cause the transition dipoles to tilt towards the fibre axis with a corresponding decrease in the fluorescence anisotropy

11.11   Future developments

11.11.1   New chromophores for two-photon fluorescence

Applications of two-photon fluorescence in biology have to date largely relied upon the use of standard single-photon fluorophores, which are not necessarily the optimum candidates for two-photon excited fluorescence. Theoretical models of two-photon absorption in polyenes indicate that an elongated conjugated system coupled with strong electroactive end groups (electron donating or withdrawing) will significantly enhance the two-photon absorption cross-section. In this respect two-photon quadrupolar systems have proved most successful. The design utilizes a centro-symmetric system of donor and acceptor groups connected by a conjugated π-electron system; compounds based on photo-stable fluorescent cores such as biphenyl and fluorene have proved most successful, combining a large two-photon cross section with a high quantum yield of fluorescence. Two examples of such compounds, OM62 and LP79, developed at the University of Rennes, are illustrated in Figure 11.20; their photophysical characteristics are listed in Table 11.2 and compared with those of the conventional fluorophore rhodamine 6G. From Equation (11.4) it can be seen that the energy (power) dependence of the two-photon transition probability scales as the square root of $\sigma^{(2)}$. As a result, LP79 and OM62 respectively require 14.5 and 33.3 per-cent of the power required for an equivalent degree of excitation in rhodamine 6G, permitting further reductions in sample irradiation with a consequent reduction in photo-induced damage. The development of such chromophores allows the possibility of new nonlinear spectroscopic techniques involving two-photon excitation to be employed without the expense and complexity of amplified ultrafast lasers.

11.11.2   Stimulated emission depletion in two-photon excited states

One of the principal and until recently apparently insurmountable barriers in far-field microscopy has been the intrinsic diffraction resolution limit below which it is impossible to distinguish between two objects. This intrinsic property of all (linear) far-field microscopy techniques was recognized by Ernst Abbe at the end of the nineteenth century. In simple terms Abbe's criterion states that it is not possible to distinguish between two objects that are closer than about one-third the wavelength of light. Recent developments in nonlinear fluorescence microscopy involving the tailoring of excited state populations by selective optical depletion have shown that this limit can indeed be broken. The technique is based on combining fluorescence microscopy with stimulated emission depletion. The initial fluorescent population is prepared by a 'pump' pulse and depletion (the return of photoexcited molecules to the ground electronic state is achieved by a second time-delayed and red-shifted 'dump' pulse, illustrated

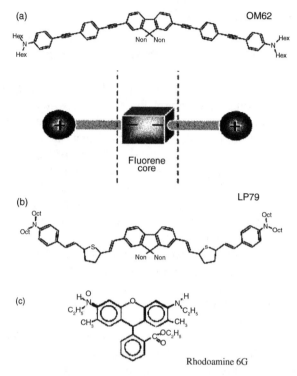

Figure 11.20   Structures of the quadrupolar two-photon polyenes OM62 (a) and LP79 (b) are engineered to undergo a quadrupolar intramolecular charge redistribution upon two-photon excitation and are based around a fluorene core with electron donating end groups. (c) A standard single-photon visible fluorescent probe rhodamine 6G

Table 11.2   Photophysical parameters for two quadrupolar push–push polyenes (LP79 and OM62) compared with those of rhodamine 6G

Fluorophore	OM62	LP79	Rhodamine 6G
Molecular Weight	1185.876	1318.208	479.02
Single Photon Absorption Maximum (nm)	387	470	528
$\sigma (\times 10^{-6}\ cm^2)$	2.13	2.04	1.93
Emission Maxima (nm)	433.56	524.6	555
Quantum Yield	0.82	0.47	0.95
Fluorescence Lifetime (ns)	0.692	0.869	5.7
	(in toluene)	(in toluene)	(ethylene glycol)
TPA Maximum (nm)	705	705	700
$\sigma^{(2)}$ at TPA Maximum $(\times 10^{-48} cm^4\ s/photon)$	10.75	54.85	1.16
$\sigma^{(2)}$ at 800 nm $(\times 10^{-48} cm^4\ s/photon)$	4.39	17.73	0.36

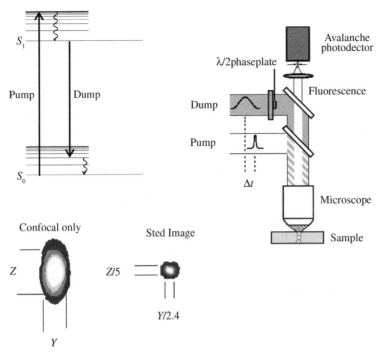

Figure 11.21 Stimulated emission depletion (STED) microscopy. The sample is excited using single-photon excitation (PUMP pulse) in a confocal microscope arrangement. A time-delayed DUMP pulse selectively depletes close to 100% of the excited state population in a region around the focus of the PUMP pulse. Using this approach, Hell and co-workers were able to obtain a 5-fold reduction in the fluorescent spot size in the vertical (Z-direction) and a greater than a 2-fold reduction in the horizontal (Y/X) direction, leading to a final image size of 97 by 104 nm

figuratively in Figure 11.21. If the pump and dump pulses are spatially displaced, it is possible to form a highly localized fluorescent region in the sample. This effectively restricts the volume from which spontaneous emission occurs and, as a result (with the appropriate spatial and temporal shaping of the dump pulse), it is possible to obtain spatial resolution that breaks Abbe's criterion. Stimulated emission depletion (STED) microscopy has to date been limited to single-photon excitation. STED has recently been demonstrated in two-photon excited states, as shown in Figure 11.22. With advances in laser sources and the development of molecular probes optimized for both two-photon excitation and STED, further improvements three-dimensional spatial resolution look promising. In addition it has been recently shown that STED is also of considerable value as a spectroscopic technique; in particular it permits the measurement of higher-order parameters for *excited* states, in particular the $\langle \alpha_{40} \rangle$ moments introduced in Equation (11.30) (Figure 11.17), something that is forbidden in conventional fluorescence techniques.

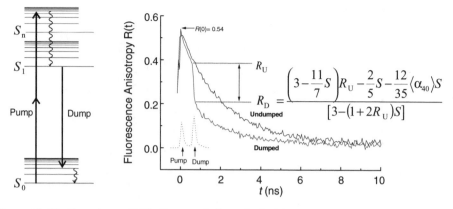

Figure 11.22 Two-photon STED. The sample is excited with a ca. 100 fs laser pulse (PUMP). A time-delayed picosecond laser pulse (DUMP) is used to partially deplete the excited state population. This causes an abrupt change in fluorescence polarization due to the preferential removal of molecules oriented parallel to the polarization direction of the PUMP. Analysis of the anisotropy change shows that it is sensitive to both the initial (pre-dumped) fluorescence anisotropy (R_U) and the pre-dumped excited state value of $\langle \alpha_{40} \rangle$. This order parameter is unobservable in conventional fluorescence experiments

11.12 Conclusions

Single- and two-photon fluorescence techniques have permitted significant advances in imaging and understanding the interactions between molecules in cells (see sections 11.5 and 11.7.2). Two-photon fluorescence offers the biologist a number of distinct benefits; firstly in its contribution to 'conventional' fluorescence imaging where it provides comparable three-dimensional resolution to scanning confocal microscopy but is far less invasive. As fluorescence is necessarily confined to the focal region of the laser the drawbacks of widespread sample illumination (photo-induced damage) are significantly reduced. Two-photon fluorescence analogues of conventional single photon techniques (e.g. fluorescence lifetime and anisotropy measurements) are becoming widely used. With the development of new two-photon fluorophores together with precision imaging and new spectroscopic techniques such as STED it is clear that further advances in biological imaging and spectroscopy can be expected.

Further reading

Fluorescence Spectroscopy

Lakowicz, J. R. (1999). *Principles of Fluorescence Spectroscopy*. 2nd Ed Kluwer Academic/Plenum Publishers, New York.
Valeur, B. (2002). *Molecular Fluorescence Principles and Applications*. Wiley-VCH: New York.

Fluoresecence Lifetime Imaging and RET

Calleja, V., Ameer-Beg, S.M., Vojnovic, B., Woscholski, R., Downward, J. and Larijani, B. (2003). Monitoring conformational changes of proteins in cells by fluorescence lifetime imaging microscopy. *Biochem. J.* **372**: 33–40.

Larijani, B., Allen-Baume V., Morgan, C.P., Li, M. and Cockcroft, S. (2003). EGF regulation of PITP dynamics is blocked by inhibitors of phospholipase C and of the Ras-MAP kinase pathway. *Curr. Biol.*, **13**: 78–84.

Two-photon absorption

Loudon, R. (1988). *The Quantum Theory of Light*, 2nd edn. Oxford University Press: Oxford.

Wan, C. and Johnson, C.K. (1994). Time-resolved anisotropic two-photon spectroscopy. *Chem. Phys.* **179**: 513–531.

Two-photon fluorescence microscopy

Denk W., Strickler, J.H. and Webb, W.W. (1990). Two-photon laser scanning fluorescence microscopy. *Science* **248**: 73–76.

So, P.T.C., Dong, C.Y., Masters, B.R. and Berland, K.M. (2000). Two-photon excitation fluorescence microscopy. *A. Rev. Biomed. Engng.* **2**: 399–429.

Squier, J. and Muller, M. (2001). High resolution nonlinear microscopy: a review of sources and methods for achieving optimal imaging. *Rev. Sci. Instrum.* **72**: 2855–2867.

Fluorescence polarization techniques and ordered molecular systems

Bain, A.J. (2002). Time resolved polarized fluorescence studies of ordered molecular systems. In *An Introduction to Laser Spectroscopy*, Andrews, D.L. and Demidov, A., eds, Chapter 6, pp. 171–210. Kluwer Scientific: London.

Bain, A.J., Chandna, P., Butcher, G. and Bryant, J. (200). Picosecond polarized fluorescence studies of anisotropic fluid media I and II. *J. Chem. Phys.* **112**: 10418–10449.

Chadborn, N., Bryant, J., Bain, A.J. and O'Shea, P. (1999). Ligand dependent conformational equilibria of serum albumin revealed by tryptophan fluorescence quenching. *Biophys. J.* **76**: 2198–2207.

Hopkins, S.C., Sabido-David, C., Corrie, J.E., Irving, M. and Goldman, Y.E. (1998). Fluorescence polarization transients from rhodamine isomers on the myosin regulatory light chain in skeletal muscle fibers. *Biophys. J.* **74**: 3093–3110.

New fluorophores for two-photon absorption

Marsh, R.J., Leonczek, N.D., Armoogum, D.A., Porres, L., Mongin, O., Blanchard-Desce, M. and Bain, A. J. (2004). *Proc. Int. Soc. Opt. Engng.* **5510**: 117–128.

Mongin, O., Porrès, L., Moreaux, L., Mertz, J. C. and Blanchard-Desce, M. (2002). Synthesis and photophysical properties of new conjugated fluorophores designed for two-photon excited fuorescence. *Org. Letts* 719–722.

Stimulated emission depletion

Bain, A.J., Marsh, R.J., Armoogum, D.A., Porrès, L., Mongin, O. and Blanchard-Desce, M. (2003). Time resolved stimulated emission depletion in two-photon excited states. *BioChem. Soc. Trans.* **31**: 1047–1051.

Klar, T.A., Jakobs, S., Dyba, M., Egner, A., Hell, S.W. (2000). Fluorescence microscopy with diffraction resolution barrier broken by stimulated emission. *Proc. Natl. Acad. Sci. USA* **97**: 8206–8210.

Marsh, R.J., Armoogum D.A. and Bain, A.J. (2002). Stimulated emission depletion of two photon excited states. *Chem. Phys. Lett.* **366**: 398–405.

12

Optical tweezers

Christopher Batters and **Justin E. Molloy**

Division of Physical Biochemistry, MRC National Institute for Medical Research, The Ridgeway, Mill Hill, London NW7 1AA, UK

12.1   Introduction

12.1.1   History of optical tweezers

In 1970, Ashkin demonstrated the use of radiation pressure to manipulate micron-sized particles. He found that an unfocused laser beam pulled objects of high refractive index towards the centre of the beam and propelled them in the direction of light propagation; and that two counter-propagating beams allowed objects to be trapped in three dimensions. The application of radiation pressure to optically confine single atoms found immediate application in basic physics for so-called 'cold atom' studies. However, it was the later discovery that a single, tightly focussed laser beam (termed 'optical tweezer') could be used to capture and manipulate live bacteria in aqueous solution that paved the way for a host of new biological experiments. The purpose of this chapter is to introduce the reader to the technical aspects of optical tweezer construction and to show how an optical tweezers-based mechanical transducer can be used to measure the forces and movements produced by interacting biological molecules. We have assumed that most readers will have a background in biology rather than physics and hope the text is accessible and useful to the general scientific audience.

The major impact of optical tweezers in biology arises from their use in the rapidly expanding field of single-molecule research. By chance, the optical forces produced by commonly available lasers (with output powers of a few hundred milliwatts) happen to be in the piconewton (pN) range, which is just right to experiment with

Chemical Biology   Edited by Banafshé Larijani, Colin. A. Rosser and Rudiger Woscholski

biological molecules. For instance, a force of 1000 pN will break a covalent bond, 150 pN will disrupt avidin–biotin and antibody–antigen interactions, 60 pN will unravel the DNA double helix and completely unfold proteins such as titin; meanwhile, DNA processing enzymes can produce about 30 pN of force and muscle proteins produce about 10 pN; finally, weak, protein–protein interactions are in the sub-piconewton range. In single molecule mechanical experiments, laser power is simply 'dialled up' to produce the right amount of force. Critical to these studies is the ability to measure nanometre-sized displacements and to calibrate the applied optical forces. This requires an optical tweezer apparatus or transducer to be constructed. Crude optical tweezer set-ups are now commercially available. However, most high-precision measurements are made using custom-built equipment because each new experimental study generally has its own particular nuance that requires modifications to be made to the apparatus. With computer-control comes versatility and the system we describe here has the advantage that many new experiments can be devised and performed simply by making modifications to computer software.

12.1.2 Single-molecule studies

Most of our current understanding of chemical reactions comes from experiments made on ensembles of molecules. However, these experiments report only average properties and often we would like to know much more about the detailed molecular mechanisms. Studying single molecules can give unequivocal insights into enzyme mechanism and these experiments have only recently been made possible by the arrival of new technologies. The first single-molecule studies in biology were single-ion channel patch-clamp recordings made by Sakmann and Neher (1976). With the development of atomic force microscopy (AFM) and optical tweezers we have been able to measure the mechanical properties of individual biological molecules. Finally, the recent development of single-molecule imaging methods using total internal reflection fluorescence microscopy (TIRFM) and single-particle electron microscopy has enabled observation of individual molecules both *in vitro* (either isolated on a microscope coverslip or embedded in vitreous ice) and in living systems. Many single molecule studies have focussed on motor proteins, which have become the paradigm or prototype system to study.

Optical tweezers-based, single-molecule mechanical studies allow single-enzyme turnovers to be observed directly and the appeal of studying energy transducing enzymes, like motor proteins, lies in uncovering details of the coupling between chemical and mechanical change. The conversion of the free energy of ATP hydrolysis to mechanical work by motor proteins occurs in several discrete phases and these phases can be probed during an individual chemo-mechanical turnover. Unlike bulk measurements there is no need to synchronize a population of molecules because transient intermediates and even fluctuations between equilibrium states can be directly observed. Also, heterogeneous behaviour of different molecules can be studied along with rare events, which would normally be masked by bulk measurement methods. Finally, the genome projects have revealed that the prototype systems that we have been studying for many years (e.g. myosin from muscle and sodium channels in nerve) are in fact members of large, homologous, protein families (i.e. in humans there are 39 myosin and >9 sodium channel genes). Many of these gene products are present in only very small numbers of molecules per living cell and some actually perform their physiological function by working as independent molecules. This gives new and increased relevance to single-molecule experimental studies in terms of providing both the necessary tools to study isolated molecules and giving detailed mechanistic insights that permeate across all levels of cell biology and systems biology.

12.2   Theoretical background

Every photon of light carries energy $\hbar\omega$ and momentum $\hbar k$ so, if light is *absorbed* by an object, the momentum transferred from a light beam of power, P, leads to a reaction force, F, on the object, given by

$$F = \frac{nP}{c}$$

where c is the velocity of light *in vaccuo* and n is the refractive index of the surrounding medium (Molloy and Padgett, 2002). A light beam with normal incidence to a *reflective* surface is returned along its own path so that the transfer of momentum is twice the above value. Transfer of momentum due to absorption and reflection (*scattering force*) is in the direction of propagation of the incident beam (i.e. acts to push the object along the light path). However, light that is *refracted* by an object that has a higher refractive index than the surrounding medium gives a transfer of momentum with a vector that points back towards the direction of the incoming beam (i.e. pulling the object towards the light source). These forces are known as *gradient forces* because for small specimens they are proportional to the gradient in light intensity (see Box 1, left panel).

For transparent objects that are of the same order of size as the wavelength of light, the net transfer of momentum depends on a delicate balance between gradient and scattering forces and is a fairly complex function of wavelength, optical intensity gradient, particle size, shape, refractive index and polarizability. The important outcome of the detailed theory is that, when an extremely high gradient of light intensity is produced by a high numerical aperture microscope objective lens, gradient forces dominate and transparent micron-sized particles of high refractive index are attracted to a stable position close to the laser focus. Particles become trapped or held in three dimensions and experience a restoring force that is proportional to displacement over a range of about 500 nm, and this linear stiffness is directly proportional to optical flux (i.e. laser input power). Beyond about 500 nm from the trap centre particles enter an unstable region, where scattering forces start to dominate and they are then pushed out of the light beam (Box 1).

Box 1   How optical forces are generated in optical tweezers

The upper panel shows how the gradient in light intensity across a single mode Gaussian laser beam (image in the centre of the panel) produces a lateral restoring force (left) and a three-dimensional single beam, gradient trap, when a tightly focused beam is used (right). The lower panel shows that rays with a high angle of incidence (>30°) produce a net force that acts towards the direction of light propagation.

Lateral trapping force   Axial trapping force

Laser spot

Intensity

Reaction force on particle

Transverse momentum imparted to light

Intensity

Reaction force on particle

Axial momentum imparted to light

Net, z-axis, trapping force

Angle of incidence

Reflection 4%

Absorption

Refraction 96%

This trapping geometry is called a *single beam gradient trap* or *optical tweezers*. The term tweezers is somewhat misleading as it gives the impression that two beams are required. In fact, just a single beam of light is used, but it is the peripheral optical rays emanating from the outer edge of the objective lens that provide the all-important z-axis stability (i.e. the attractive force that pulls the bead into the focus). For this reason, it is important to overfill the back aperture of the objective lens to ensure that there is a sufficient flux of these peripheral light rays. To maximise the tweezer stiffness the laser focus should be the same size as the object so that nearly all the light passes through it. The flux of light is then extremely high and to minimize heating due to light absorption a near-infrared laser (e.g. 1064 nm) is often used. The transparency of the object used (polystyrene or glass) and its high rate of heat transfer to the bathing medium (water) mean that heating is rarely a problem.

12.3   Apparatus

12.3.1   Building an optical tweezers transducer

The optical tweezer transducer shown in Figures 12.1 and 12.2 is based around an inverted optical microscope (Zeiss Axiovert series). One of the most critical components in the system is the high numerical aperture (NA) objective lens used to create the optical tweezer. Most high magnification, oil immersion, objective lenses have a sufficiently large NA (usually 1.3) and we have found that fairly inexpensive lenses trap well (e.g. Acroplan and Acrostigmat Zeiss lenses). Spherical aberration is the most important lens defect that impairs trapping efficiency and the more expensive lenses may offer important advantages in terms of spherical aberration. However, some (e.g. Zeiss Plan Apochromat series) have rather poor transmission in the near-infrared (1064 nm) and we find they do not trap well. Water immersion objective lenses are worth considering, as their spherical aberration does not change when the lens is focussed through the aqueous solution. This means that one can trap at much greater depths within the sample chamber (e.g. up to 150 μm sample penetration vs 20 μm for most oil immersion lenses). In order to make calibrated, single-molecule, mechanical measurements it is essential to mount the apparatus on a vibration-isolated table. There are many possibilities here and choice of table depends upon how noisy the work environment is (e.g. local vibration caused by machinery, people walking along corridors, etc.), available space and budget. We have found relatively low specification Workstation (Newport Corp, Irvine, CA, USA) or Lab Table (TMC, Peabody, MA, USA) to be sufficient.

To produce the laser tweezer, the best current option is to use a diode-pumped, solid-state (DPSS), continuous wave (cw), single-Gaussian transverse emission mode (TEM 0,0), near-infrared laser. The laser material is usually Nd–YAG, or similar, and this emits laser light at around 1064 nm and up to 10 W power. The jargon terms used are: 'DPSS, cw, TEM 0,0, Nd–YAG'. An Internet search will show that there is a bewildering choice of designs and manufacturers. The technology changes fast, so advice here does not help much. When budgetting for this expensive item, you should

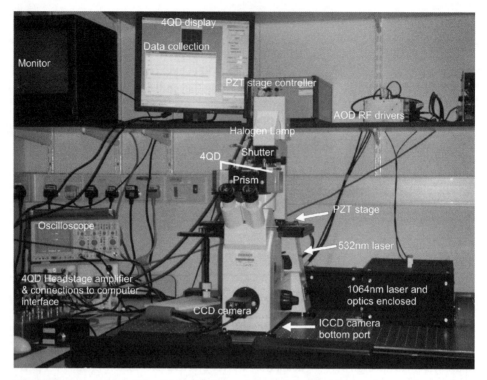

Figure 12.1 An optical tweezers-transducer. Based around an inverted microscope, all the major components are highlighted. The optical path of the ND-YAG (1064) trapping laser (class 3B) are enclosed for safety reasons. The Zeiss Axiovert and other components are mounted on a vibration isolation air table. Further details are given in the schematic shown in Figure 11.2 and in the text

remember that you will probably get only 5 years service before the laser fails. For high-precision measurements, the beam-pointing stability (tendency for the beam to wander due to thermal instability) must be <50 μrad. Remember that 10 μrad of angle change means 10 μm of movement over a 1 m path length. So, given the magnification of the optical system that you will build (see Figure 12.2) this will be equivalent to around 1 nm displacement at the specimen plane.

12.3.2 Creating multiple laser traps: acousto-optical deflectors

To move the laser tweezers the laser beam can be steered using computer-controlled acousto-optic deflectors (AODs; Figure 12.3). They consist of a block of crystalline material (often tellurium oxide) which is excited by a piezoelectric device to give a travelling acoustic wave (acoustic velocity =600 ms). This acts as a variable diffraction grating whose spacing depends upon the frequency (wavelength) of the acoustic wave (30–80 MHz). This is controlled by a high power, radio frequency (RF) driver. When the AOD crystal is tilted to the Bragg angle, 80 per cent of the light that enters is reflected into the first-order diffracted beam. Two crystals combined in series with their axis orthogonal to one another give x- and

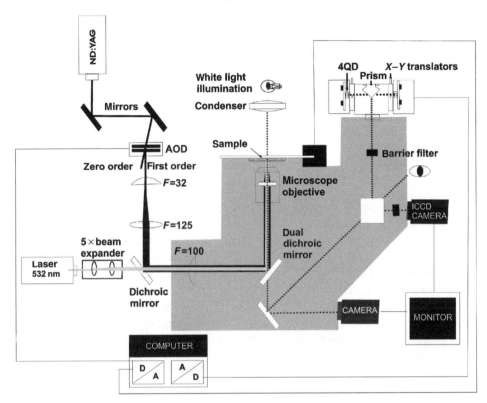

Figure 12.2 A schematic of the optical tweezers. All the components discussed in this chapter are highlighted. The solid black line represents the infrared (1064 nm) laser, the grey line represents the green (532 nm) laser and the dotted line represents white light optical paths. The grey box represents the parts enclosed in the microscope

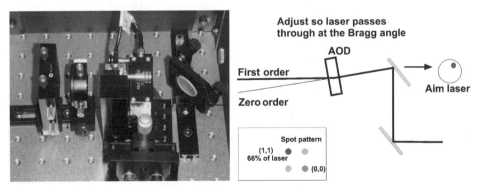

Figure 12.3 Acousto optical defelectors (AODs) are used to steer the laser optical tweezer. Left: a photograph of the AODs mounted on the optical table. The AODs are fixed to translators that allow fine adjustment of angle and translation with respect to the input laser beam. Right: how the laser must enter the AOD at the Bragg angle so that the first order beam (which is the beam used to produce the optical trap) exits aligned with the microscope axis

y-axis control. The overall efficiency will then be around 64 per cent. The diffraction angle is controllable over a range of about 30 mrad and can be changed within about 2 μs. This means that multiple optical tweezers can be produced by rapid scanning of a single laser beam between two or more trap positions (time-sharing). This time-sharing approach works because viscous drag on the trapped objects is high enough to provide positional 'persistence' while the laser beam is elsewhere. Computer control means that the tweezer can be moved with high precision in a variety of pre-programmed waveforms.

12.3.3   Camera and light path

Use of a research-quality fluorescence optical microscope as the basis for building an optical tweezer has the advantage that camera ports, filter holders, mechanical stage, objective lens focussing, condenser illumination and so on are all accurately made and aligned, well tested and ergonomically designed. The only modifications required to start trapping are: (1) insertion of a suitable dichroic mirror into the fluorescence filter holder so that laser light is reflected onto the specimen (a dual dichroic will reflect 532 and 1064 nm light but transmit 700 nm red light; Omega Optical Brattleboro, Vt, USA); (2) construction of a simple adaptor to allow the infrared laser light to be introduced into the epifluorescence light path at the back of the microscope (Figure 12.4)

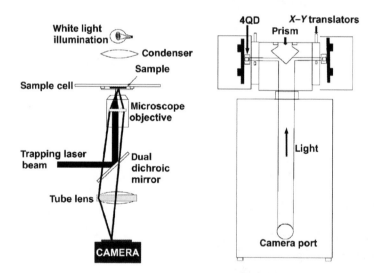

Figure 12.4   Visualizing and monitoring bead position. Left: where the dual dichroic mirror is positioned in the microscope. This allows the 532 and 1064 nm laser light to be reflected into the sample cell. Red (590→720 nm) light is transmitted through the mirror so that rhodamine fluorescence can be visualized using an image intensified camera system. Right: diagram of the microscope and the four-quadrant photodiode detectors (4QD). Light can enter the camera port for visualization or can be directed to the 4QDs for data collection. The Amici prism is front-surface mirrored and splits the light path into two halves, allowing both beads in the three-bead trapping geometry to be monitored (see text for details). The translators allow independent fine positioning of the quadrant detectors so the images of the beads are always in the centre of each 4QD

In order to make calibrated measurements from single molecules two more items are required: (1) a four-quadrant photodiode position sensor (4QD) to accurately measure the position of the bead held in the optical tweezer (Figure 12.5); (2) a piezoelectric substage to enable controlled movements to the specimen during the experiments and in order to calibrate the system. Both of these components can be interfaced to the computer using A/D and D/A convertors.

12.4 Data collection and analysis

12.4.1 Collecting data with an optical tweezer

Nearly all optical tweezer experiments involve the use of micron-sized polystyrene beads, which act as handles to manipulate the proteins under investigation. The use of beads provides a highly versatile system; they are available in a wide variety of sizes, colours and surface chemistries. One-micrometre beads are very stably trapped using 1064 nm laser light and also give a high-contrast image that can be cast onto the 4QD, creating a nanometre resolution position sensor simply using bright field optical microscopy (Figure 12.5).

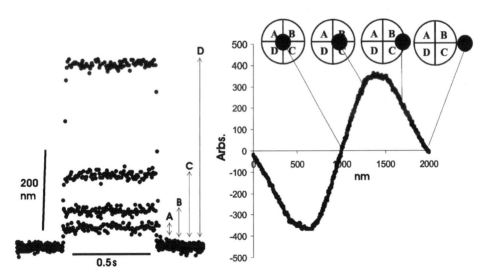

Figure 12.5 Position calibration. Left: a bead is held in the optical tweezers and a 1 Hz squarewave oscillation is applied to the AODs; the displacements are ± (a) 50, (b) 100, (c) 200 and (d) 500 nm. Right: a large amplitude triangle wave is applied to the AODs to move the bead held in the optical tweezer over the full active area of the detector. The sensitivity is greatest when the bead is in the centre of the 4QD. Ideally the image of the bead held in the optical tweezers should be approximately half the size of the 4QD. This gives greatest sensitivity and sufficient detector range to perform experiments

12.4.2 Calibration of the detectors and of the optical tweezer stiffness

Before experiments are performed it is necessary to calibrate the position sensitive 4QD detector and the stiffness of the tweezer. First, the AODs are calibrated against a microscope graticule by measuring the input signal required to move the laser beam by a known amount. This involves use of a video camera system to image the graticule and laser spot simultaneously. To determine the position sensitivity, a square wave of known amplitude is applied using the AODs to deflect the laser beam while a 1 μm bead is held in the optical tweezers, with its image centred on the 4QD (Figure 12.5, left). Alternatively, the bead can be scanned using a triangle wave displacement to give a continuous calibration of position across the entire active area of the detector (Figure 12.5, right)

The stiffness of the laser tweezers can be calibrated using two different methods described below. Both stiffness and position sensitivity must be calibrated for each bead used as there is considerable variability (10–20 per cent) between beads.

12.4.3 Stokes calibration

Stokes calibration involves applying a known viscous drag force to a bead held in the optical tweezer and recording how far it is displaced from the tweezer centre. Application of a triangle wave oscillation of known size and frequency to the specimen chamber with the piezoelectric substage produces a viscous drag force given by Stoke's law

$$F = 6\pi\eta r v$$

where η = viscosity, r = radius of the bead and v = velocity of the stage. By measuring the displacement of the bead (X), the stiffness of the trap can be determined (Figure 12.6)

$$\frac{\text{Force}}{\text{Displacement}} = \kappa$$

12.4.4 Equipartition principle

Brownian movement of the bead held in the optical tweezers serves as a useful probe of the tweezer stiffness. The variance of bead position along one axis, $\langle x \rangle^2$, is inversely proportional to the stiffness of the optical tweezers κ_x and is directly proportional to absolute temperature.

$$\langle x \rangle^2 = \frac{\kappa T}{\kappa}$$

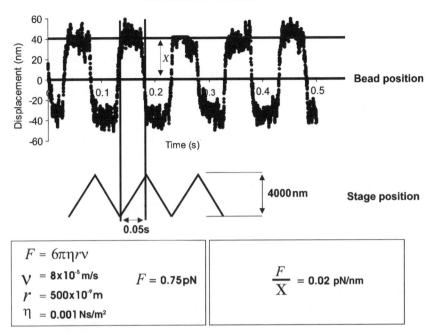

Figure 12.6 Calibration of trap stiffness using Stokes' drag. The stiffness of the trap can be determined using Stokes' drag force. A triangle wave is applied to the stage which holds the specimen chamber while a bead is held in the optical tweezer. The rapid movement of the stage creates a force, F, on the bead caused by the motion of the surrounding fluid. This causes the bead to be displaced a distance, x, from the trap centre; the greater the system stiffness the less the bead is displaced. F/X = stiffness of the tweezers (K) (inset)

Tweezer stiffness can therefore be calibrated by measuring the mean squared Brownian motion. In addition, if the power spectrum of the movement is measured we find that, because of viscous damping, thermal noise shows a Lorenzian distribution with a 'roll-off' given by

$$f_c = \frac{\kappa_x}{12\pi^2\eta r}$$

12.5   A biological application

12.5.1   Studying acto-myosin interactions

The ability to measure nanometre displacements and piconewton forces means mechanical and kinetic properties of single motor proteins can be measured. In order to study intermittently interacting motors like acto-myosin-II from muscle, a special tweezer geometry is used; a single actin filament is held taut between two beads about 5 μm apart in separate optical tweezers. The filament is then manipulated into the vicinity of a third, surface-bound bead, on which myosin molecules are present at a low surface density. The two beads (handles) are then imaged onto the quadrant detectors and the periodic binding between myosin and actin can be observed and recorded (Box 2).

Box 2   Single molecule mechanical studies of acto-myosin interactions using optical tweezers

(a) The three-bead geometry used to study intermittently interacting molecular motors like myosin-II from muscle. (b) Bright-field image of the two optically trapped beads attached at either end of a single actin filament. This image is cast onto two 4QD detectors (see text). (c) Position data plotted against time. The noisy record is due to thermal vibration of the optically trapped bead. Sudden changes in noise and displacement from the mean rest position are due to individual myosin binding events. Events are 'picked' from the raw data on the basis of changes in signal variance (lower trace), which report changes in system stiffness.

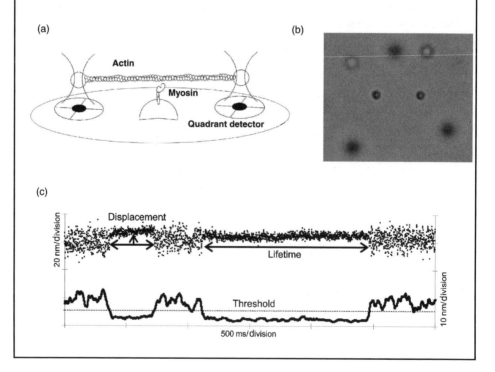

(a)

(b)

(c)

The optical tweezers stiffness is set to about 0.02 pN nm (adjusting the laser power), which is about 10 000 times less stiff than an AFM cantilever. This is sufficient to hold the beads stably trapped but still about 10→100 times less stiff than the acto-myosin complex. This means that myosin can undergo its conformational change essentially unhindered by the tweezer.

12.5.2   Identification of events and data analysis

Individual binding events between myosin and actin can be identified by monitoring the variance of the data. Myosin binding increases the system stiffness, as it introduces an additional route to the mechanical ground (the microscope coverslip). This means that the Brownian motion is reduced. A typical procedure for detecting myosin binding to actin involves:

(1) Calculate the running mean of the position data using a window size of approximately 50 ms. This is then subtracted from the original position data to remove slow changes in the mean position.

(2) Calculate the running variance (over a 10 ms window size) of the above data. The variance record is then low-pass filtered by calculating its running median (25 ms window).

(3) The variance data is then thresholded at a value between 0.04 and 0.5 pN nm. The original, unprocessed positional data is then marked to indicate the start and end of each binding event based on sudden increases in system stiffness. The attached periods are then isolated from the unbound free periods and their amplitude measured relative to the local mean position.

(4) The duration of each bound event is recorded along with the duration of the adjacent detached period preceding the event.

The data can be analysed to determine, amongst other things, the average size of the myosin powerstroke and the kinetics of the unbinding process. It is also possible to inspect individual binding events more clearly to see if they have consistent features such as a two-phase powerstroke [Box 2 (c)].

12.5.3   Measuring the powerstroke

It is necessary to collect a statistically meaningful data set for two reasons:

(1) in order to determine the orientation of the actin filament, so that data can be pooled from different bead pairs and experimental runs;

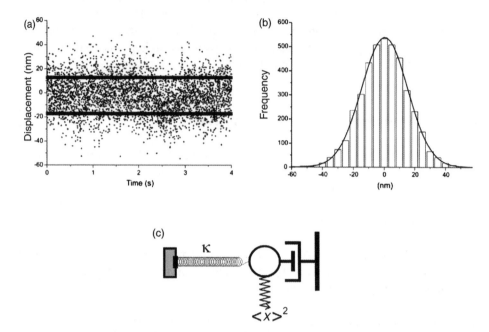

Figure 12.7 Calibration of trap stiffness by thermal noise analysis. A single 1.1 μm bead is held in the optical tweezers and data is collected at 2 kHz. (a) The graph shows the bead position vs time – solid lines denote ±1 standard deviation of bead position. (b) The same data plotted as a histogram. The mean displacement is 0 nm and the variance is determined by the vibration due to brownian noise. (c) The trap stiffness can be determined from this information using the equipartition principle: $1/2\kappa_{trap}\langle x^2 \rangle = 1/2k_BT$, where κ_{trap} = trap stiffness, $\langle x^2 \rangle$ = variance, k_B = Boltzman constant and T = absolute temperature

(2) the absolute accuracy of the measurement is determined by

$$\text{SEM} = \frac{\sigma}{\sqrt{n}}$$

Since most of the variance in the data arises from thermal motion of the beads held in the optical tweezers, the standard deviation (σ) is governed by the optical tweezer stiffness (Figure 12.7). Therefore, if for example experiments are carried out at a combined laser tweezer stiffness of 0.04 pN nm, the standard deviation of the thermal noise will be 15 nm, so if 100 events are collected the step size estimate will have a standard error of the mean of $(15/\sqrt{100}) = 1.5$ nm.

To check that the displacement amplitudes of the individual binding events are distributed according to the above model, in which thermal noise produces the spread in the data, a histogram is plotted to which a Gaussian fit can be applied (Figure 12.8). The step size is determined from the shift in this distribution from zero (the mean value of the noise). This is more easily understood if we consider the

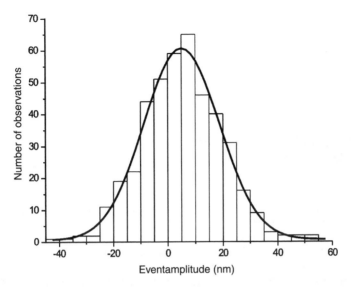

Figure 12.8 Analysis of acto-myosin binding events. Data obtained during acto-myosin interactions (see Box 2) are analysed in a way that must account for the large amount of background thermal noise. The histogram shows the distribution of event amplitudes obtained using rabbit skeletal myosin-II S1 fragment interacting with rabbit actin at room temperature and 10 μM ATP. The fitted line is a Gaussian distribution, which is identical to the distribution obtained for the bead positional noise in the absence of binding events [Figure 6(b)], except that it is shifted from zero by an amount equivalent to the myosin working stroke, in this case approximately 5.5 nm

following example: if we measure the heights of 1000 people in their socks and plot the data, we will obtain a normal distribution of heights. If this process is repeated, but this time everyone now wears their shoes, we will obtain a similar distribution but it will be shifted to the right. The size of this shift reveals the average height of the shoes.

12.5.4 Lifetime analysis

Myosin is an actin-activated ATPase which couples the hydrolysis of ATP to the production of force and movement. The ATPase cycle occurs in the following way: myosin starts its hydrolysis cycle with ATP bound in the catalytic domain. ATP is then hydrolysed to ADP and P_i. Myosin then binds to actin and the products, first phosphate and then ADP, are released. During this actin-attached part of the cycle, movement and force are produced when myosin changes conformation. Myosin remains bound to actin in a nucleotide-free 'rigor' state until a fresh ATP molecule diffuses into its catalytic site, causing it to dissociate from actin. The cycle then starts again. The bound lifetime of a single acto-myosin complex depends on the probability of that event ending. At very low concentrations of

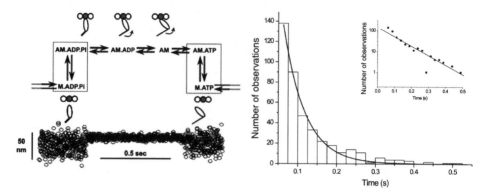

Figure 12.9 Single molecule studies can reveal enzyme kinetic information. Left: the acto-myosin ATPase cycle lined up with a typical binding event; the event begins when myosin binds to actin and ends when an ATP molecule diffuses into the binding cleft, causing it to dissociate from the actin filament. Right: the event lifetimes of rabbit skeletal myosin-II S1 fragment are stochastic and show an exponential distribution because each event is terminated by a single Poisson process (ATP binding). Note that the inset has two points removed, as zeros cannot be plotted on a log scale

ATP this is governed by the rate of ATP binding to the rigor complex. Event durations are therefore stochastic in nature and depend on a single Poisson process (ATP binding), and this gives rise to a distribution of lifetimes with a single exponential decay (Figure 12.9).

12.6   Other biological examples

Optical tweezers have been used to study a wide variety of different biological systems. One of the first studies measured the torque produced by the bacterial flagella motor. A single bacterium, tethered to a microscope coverslip by its flagellum, was held in optical tweezers and the force required to turn it either with or against the action of its flagella motor was measured. Later, the stepping behaviour of individual kinesin molecules was measured by fixing a microtubule to a coverslip and using optical tweezers to hold a kinesin-coated bead just above it. The kinesin was then able to grab onto the microtubule and tug the bead along, gradually pulling the bead out of the optical tweezer. This study proved that it was possible to resolve the nanometre-sized displacements (8 nm each) produced as each ATP molecule was broken down, causing kinesin to advance one step forward. Workers then turned their attention to the muscle protein, myosin, which proved more difficult to work with as it interacts only intermittently with the actin filament. Acto-myosin studies were dealt with earlier in this chapter.

Box 3   Optical tweezers have been used to study a wide variety of different biological systems

The uses of optical tweezers are extending beyond molecular motors; they are manifold and exhaustive.

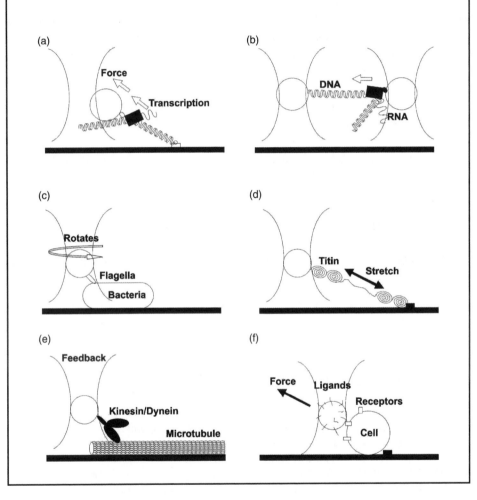

Optical tweezers have since been used to: stretch and even unwind DNA; unfold the tertiary structure of proteins; pull on DNA processing enzymes like RNA polymerase as it walks along DNA; resist the force produced by viral capsid packaging motors; and manipulate entire mammalian cells, sperms and sub-cellular organelles and vesicles. The applications are manifold and the future possibilities are great.

12.7 Summary

The purpose of this chapter has been to give the reader a practical introduction to optical tweezers; how they work, how they are built, how they can be used to make calibrated measurements on single molecules and how a paradigm system like actin and myosin can be studied. A future challenge is to combine the use of other single molecule techniques with optical tweezers so that parallel measurements can be made on the same single molecule. Such studies will give detailed insights into enzyme mechanism. One would also like to make many simultaneous measurements from multiple, individual molecules within a living cell so that we can build up a picture of how proteins, DNA and ligands interact in intact systems.

Acknowledgements

The authors are grateful for financial support provided by the Medical Research Council UK and the UK IRC in Bionanotechnology.

Further reading

Greulich, K.O. (1999). *Micromanipulation by Light in Biology and Medicine*. Birkhauser: Berlin.

Sheetz, M.P. (1998). *Laser Tweezers in Cell Biology*. Methods in Cell Biology, Vol. 55. Academic Press: San Diego, CA.

Ashkin, A. (1970). Acceleration and trapping of particles by radiation pressure. *Phys. Rev. Lett.* **24**: 156–159.

Ashkin, A., Dziedzic, J. M., Bjorkholm, J. E. and Chu, S. (1986). Observation of a single-beam gradient force optical trap for dielectric particles. *Opt. Lett.* **11**: 288–290.

Molloy, J. E. and Padgett, M. J. (2002). Lights, action: optical tweezers. *Contemp. Phys.* **43**: 241–258.

Molloy, J. E., Burns, J. E., Kendrick Jones, J., Tregear, R. T. and White, D. C. S. (1995). Movement and force produced by a single myosin head. *Nature* **378**: 209–212.

Neher, E. and Sakmann, B. (1976). Single-channel currents recorded from membrane of deservated frog muscle fibres. *Nature* **260**: 779–802.

Neuman, K. C. and Block, S. M. (2004). Optical trapping. *Rev. Sci. Instrum.* **75**: 2787–2809.

Smith, S. B., Cui, Y. J. and Bustamante, C. (1996). Overstretching B-DNA: The elastic response of individual double-stranded and single-stranded DNA molecules. *Science* **271**: 795–799.

Svoboda, K. and Block, S. M. (1994). Biological applications of optical forces. *A. Rev. Biophys. Biomol. Struct.* **23**: 247–285.

Svoboda, K., Schmidt, C. F., Schnapp, B. J. and Block, S. M. (1993). Direct observation of kinesin stepping by optical trapping interferometry. *Nature* **365**: 721–727.

Tskhovrebova, L., Trinick, J., Sleep, J. A. & Simmons, R. M. (1997). Elasticity and unfolding of single molecules of the giant muscle protein titin. *Nature* **387**: 308–312.

13

PET imaging in chemical biology

Ramón Vilar

Institució Catalana de la Reserca i Estudis Avançats (ICREA) and Institute of Chemical Research of Catalonia (ICIQ), 43007 Tarragona, Spain

13.1   Introduction

Monitoring molecular events *in vivo* and in real time is essential to gain a better fundamental understanding of biochemical processes, to develop novel means of early detection of disease and to design new drugs. Consequently, there has been great interest in finding suitable techniques with the appropriate sensitivity, selectivity and spatial resolution to be able to image – at a molecular level – biological processes. Several *in vivo* imaging methods (such as X-ray) rely exclusively on imaging gross anatomy with the restriction that only structural imaging can be carried out and no information can be obtained on metabolic and molecular events. Similarly, for the *in vivo* detection of disease, these imaging modalities are limited to those malfunctions associated with structural abnormalities.

In contrast, techniques such as optical imaging, positron emission tomography (PET) and single-photon emission computed tomography (SPECT) have the ability to monitor metabolic processes *in vivo*. These techniques make use of highly sensitive imaging agents or reporter probes that can be designed to selectively interact with specific tissues or biomolecules, providing a much more detailed picture of the targeted structure or biological processes at a molecular level.

This chapter will concentrate on PET as a molecular imaging modality. An introduction to the technique will first be provided followed by specific examples where PET has been successfully used to obtain a better understanding of specific biological processes, in the detection of disease and also as a tool for drug discovery and development.

Chemical Biology   Edited by Banafshé Larijani, Colin. A. Rosser and Rudiger Woscholski
© 2006 John Wiley & Sons, Ltd

13.2   Positron emission tomography: principles and instrumentation

PET is a noninvasive technique that uses radioisotopes as molecular probes to image biochemical processes in specific organs and it is therefore frequently used when the biological function is more important than the organ's physical structure (e.g. to study Alzheimer's disease, in oncology or to monitor malfunctions of the heart). This is one of the great advantages of PET: it provides metabolic information that cannot be generated with techniques confined to determining the physical structure of the organ (e.g. X-ray, ultrasound and magnetic resonance imaging, MRI). Furthermore, since some of the positron-emitting radionuclides are low atomic mass elements found in biomolecules (e.g. C, N and O), it is possible to label directly the biologically relevant species without interfering with their biological activity (e.g. incorporating a positron-emitting carbon isotope onto a drug to study its interaction with a specific enzyme). This is in contrast to other modalities where the imaging agent is a relatively large molecule itself – for example a gadolinium (III) complex for MR imaging – and hence when attached to the targeting species it usually modifies its bio-activity.

Box 1   What is a positron?

The existence of positron was first proposed by P.A.M. Dirac in the late 1920s and was experimentally discovered in 1932. A positron is the antimatter counterpart of an electron and hence has the same mass as the electron but opposite charge. The process by which a nucleus undergoes positron decay generates a new nucleotide with one fewer proton and one more neutron, besides emitting one positron and a neutrino.

$$A_Z X_N \rightarrow A_{Z-1} Y_{N+1} + e^+ + \upsilon$$

Similarly to what occurs to an electron, when a positron passes through matter it experiences a loss of energy through ionization and excitation of nearby molecules and atoms. At some point (typically after having travelled a distance of approximately 1 mm) the positron has lost enough energy to be annihilated by colliding with an electron close by.

$$e^+ + e^- \rightarrow \gamma + \gamma$$

As a consequence of this process, γ radiation is produced in two opposite directions and with energy of 511 keV.

Table 13.1 Most common positron emitting isotopes and their half-life

Radionuclide	Half-life (min)
^{11}C	20.40
^{13}N	9.96
^{15}O	2.07
^{18}F	109.70

The most commonly used radioisotopes for PET are listed in Table 13.1. Since their half-lives are on the order of tenths of minutes, an important challenge in PET imaging is to develop efficient and fast chemical processes to synthesize the radio-actively labelled molecules of interest. The short half-lives of the radioisotopes used in PET represent an important limitation of the technique since the labelled probes must often be prepared in a cyclotron in close proximity to the PET instrument.

Once the target molecule has been labelled, it is possible to measure the distribution of the positron-emitting probe in the organism under study and hence investigate a wide range of physiological and biochemical events in real time. PET has also been successfully employed to investigate drug–receptor and drug–enzyme interactions *in vivo*, providing a powerful tool to aid in the discovery of new drugs.

The principle behind PET is schematically shown in Figure 13.1. A molecule labelled with a radioisotope emits a positron (β) which can then collide with an electron present in molecules around the labelled species. When the positron and

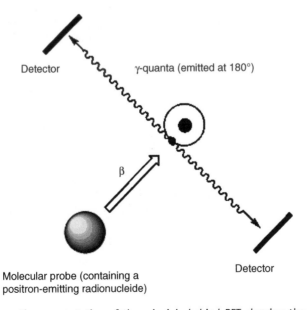

Figure 13.1 Schematic representation of the principle behind PET showing the positron decay and annihilation resulting in two 511 keV γ-quanta

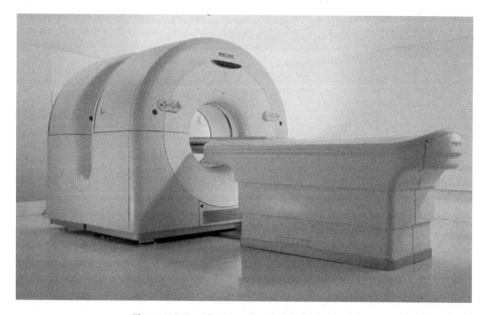

Figure 13.2   Photograph of a PET instrument

electron collide they annihilate each other, emitting γ radiation of 511 keV at 180° relative to the collision. The γ–quanta are then detected by two opposite detectors (coincidence detection) and one event (i.e. the decay of a positron) is registered by a coincidence circuit. Owing to this requirement of the technique, a PET camera is typically made from a planar ring of photon detectors, with each one of them placed in time coincidence with the detectors located on the other side of the rings (see Figure 13.2 for a photograph of the equipment used in PET imaging).

13.3   Applications of PET imaging in the biomedical sciences

The development of novel molecular probes for PET imaging requires a highly multidisciplinary approach. It involves the design and development of appropriate chemical methodologies to synthesize and purify the labelled species in a short period of time (owing to the short half-lives of the radioisotopes used). It is also necessary to have insight from physical scientists for the operation of the cyclotron to produce the labelled radioisotopes. Depending on the applications for which the labelled molecule is to be used, it is then necessary to have a specialized team of biomedical scientists to carry out the relevant experimental and clinical investigations to evaluate the imaging agent. Finally, insight from computer scientists is required to process the information collected and convert it into images that can provide the information required.

There are two general types of probes commonly used in PET imaging: direct binding probes and indirect probes. In the former, the labelled species directly binds to the targetted molecule to be imaged (e.g. a labelled species that is selectively trapped by a biological receptor). In contrast, indirect probes reflect the activity of a specific process rather than the concentration of the targeted macromolecule. Examples of both types of probes are presented in this section.

13.3.1   A labelled glucose analogue: an indirect probe to measure energy metabolism

A widely used molecular probe in PET imaging is the glucose mimic 2-deoxy-[^{18}F]fluoro-D-deoxyglucose (FDG), which has been successfully employed in neurological, cardiovascular and oncology investigations. This probe has a relatively long half-life of 109 min, which makes its production/utilization in hospitals more practical than probes based on other radionuclides with shorter half-lives.

How does FDG work? Using similar pathways to glucose, once FDG is intracellularly transported it can be phosphorylated to FDG-6-phosphate by the enzyme hexokinase (Figure 13.3). However, in contrast to phosphorylated glucose, FDG-6-phosphate is not a significant substrate for further reactions. Hence, the phosphorylated FDG is accumulated in cells, reflecting hexokinase activity and providing regional information on energy metabolism. Since FDG gives information on the activity of the enzyme and not the concentration of the specific macromolecule, it can be categorized as an indirect imaging probe.

The information obtained can then be employed to identify and characterize diseases related to alterations in the metabolism of glucose. For example, it has recently been reported that, using this approach, it is possible to detect Alzheimer's disease with more than 90 per cent accuracy three years earlier than conventional clinical diagnosis methods. On the other hand, experimental and clinical studies have shown that the uptake of FDG in cancer cells (with increased levels of glycolysis) correlates with the tumour growth rate, the degree of metastasis and the number of viable tumour cells. Consequently, radiolabelled FDG has proven to be a powerful imaging agent for locating a tumour, determining whether or not it is malignant, establishing whether it has spread and also evaluating whether a given therapy is working.

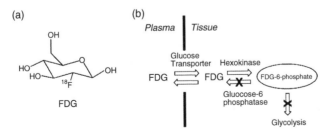

Figure 13.3   Structure of 2-deoxy-[^{18}F]fluoro-D-deoxyglucose and schematic representation of its mode of action

Box 2   Molecular imaging: comparison of the different techniques

Imaging techniques such as MRI (magnetic resonance imaging), PET (positron emission tomography), SPECT (single photon emission tomography), ultrasound and optical imaging (e.g. fluorescence-mediated molecular tomography, FMT) can be employed to analyse molecular events *in vivo* and noninvasively. These techniques should not be viewed as competing but rather as complementary since each of them has its own advantages and limitations. Among the parameters that need to be taken into account are sensitivity, spatial resolution, time-scale and cost. Depending on the problem that is being addressed, these parameters need to be considered to choose the best technique. The following table shows a comparison of some of these parameters:

Technique	Resolution	Depth	Time
MRI	10–100 µm	No limit	Minutes to hours
Ultrasound	50 µm	Millimetres	Minutes
PET	1–2 mm	No limit	Minutes
SPECT	1–2 mm	No limit	Minutes
FMT	1 mm	<10 cm	Seconds to minutes

One of the most important advantages of PET in comparison to other techniques is its extraordinary sensitivity. In contrast, it has relatively poor spatial resolution and the short half-lives of positron emitters makes it an expensive technique since an in-house cyclotron is required.

13.3.2 Imaging dopamine metabolism with PET

Dopamine is a chemical messenger produced within the nerve cells that is essential for the transmission of the nerve impulse and hence involved in a wide range of important functions, including movement, cognition and behaviour. Dysfunctions in the central nervous dopamine system can lead to diseases such as Parkinson's and schizophrenia. Alterations in the levels of this neurotransmitter have also been implicated in a variety of behavioural problems such as attention deficit and hyperactivity.

As a consequence of the important role it plays in the central nervous system, there has been very much interest in finding ways of monitoring the levels of dopamine and studying its metabolism *in vivo*. Molecular imaging techniques – PET in particular – have been increasingly useful in gaining a better understanding of these processes.

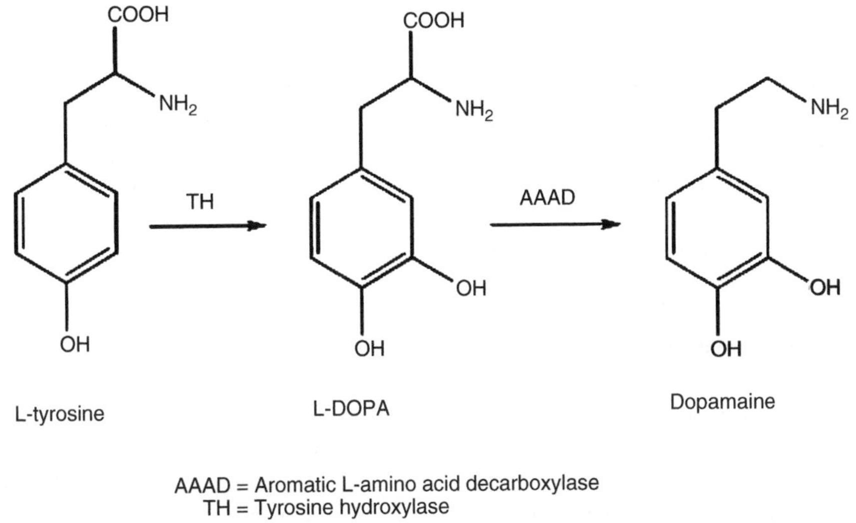

AAAD = Aromatic L-amino acid decarboxylase
TH = Tyrosine hydroxylase

Figure 13.4 Biosynthesis of dopamine from the essential amino acid L-tyrosine and schematic representation of its mode of action

The biosynthesis of dopamine starts by converting L-tyrosine to L-DOPA, which is rapidly transformed into dopamine by decarboxylation (Figure 13.4). Consequently, an obvious way to asses the levels of dopamine synthesis is by labelling its precursor and analysing its metabolism. In an early study [^{11}C]-L-DOPA with the label in the carboxylic position was prepared as a radiotracer. When this species is decarboxylated the labelled carbon is lost as $^{11}CO_2$ and as a consequence the areas where decarboxylation has occurred (i.e. where dopamine is synthesized) appear as 'cold spots' in the PET image. An improvement to this approach was made some years later by preparing [β-^{11}C]-L-DOPA where the label is located in the β-carbon position and hence is carried onto dopamine. This radiotracer has been used to understand several mechanisms taking place in the central nervous system. It has also been valuable in determining the role played by dopamine in a variety of neurological disorders. For example, by using [β-^{11}C]-L-DOPA as a PET radiotracer it has been possible to establish that dopamine synthesis in schizophrenic brains is higher compared with in normal brains.

Besides monitoring dopamine's metabolism, there is also increasing interest in finding appropriate direct probes to image its receptors in living organisms. This can be achieved by synthesizing positron-emitting analogues of dopamine (agonists) and monitoring their interactions with the corresponding receptors *in vivo*. There are now several of these species reported which have helped understanding of the different receptors associated with dopamine's function. For example 3-(2'-[^{18}F]fluoroethyl)spiperone is a ligand that binds to the dopamine D2 receptor and hence accumulates in tissues with high concentrations of these receptors. This radiotracer has been successfully employed to study the dopamine receptors of the striatum – a subcortical part of the brain associated with a variety of cognitive processes, including aspects of motor control.

These examples show how PET can provide valuable information on the metabolism and the molecular events associated with the function of an important neurotransmitter. Besides providing a better molecular understanding of the central nervous system, these radiotracers can also be of great help at the time of designing and developing drugs to treat diseases associated with dysfunctions in the central nervous dopamine system (see Section 13.3.4).

13.3.3   PET imaging in gene expression

The completion of the first draft of the genome and the recent developments in proteomics have generated a plethora of new targets for molecular imaging. Already, several imaging techniques such as MRI, PET, SPECT and optical imaging have been employed to monitor the behaviour of genes *in vivo*. However there is still a need to design and develop novel molecular probes with better sensitivity and selectivity. There is also increasing interest in multimodal probes that can make simultaneous use of the different imaging techniques to monitor gene expression *in vivo*.

There are two main approaches to imaging gene expression: by directing the studies to transgenes (i.e. genes that are externally transferred into cells or tissues) or by targeting endogenous genes. The former studies promise to be particularly valuable in monitoring gene therapy, while the latter can be very useful in gaining a better understanding of the expression of genes during development, aging and as a function of responses to external stimuli.

It is worthwhile mentioning that imaging of gene expression has actually been carried out for a while, since every time that a protein is imaged (e.g. an enzyme or a receptor) the expression of a specific set of genes is indirectly being analysed. However, the current interest has centred on imaging cellular processes that are closer to transcription (i.e. from DNA to RNA) rather than those closer to translation (i.e. from RNA to protein). This trend clearly indicates the great interest there is in imaging gene therapy and stem cell therapy.

A general (indirect) imaging approach for gene expression is by using 'reporter genes', which can track the expression of both endogenous and exogenous genes. These reporter genes can be classified into two broad categories: reporter genes that lead to the production of an enzyme that is capable of metabolizing and trapping a probe (such as a PET-labelled compound); or reporter genes that lead to the production of a protein receptor that can selectively interact with a probe. A well-studied example of a 'reporter gene' that produces an enzyme capable of metabolizing and trapping a probe is the Herpes Simplex virus type 1 thymidine kinase gene (*HSV1-tk*). This gene can be introduced into the cell using one of several potential vehicles (e.g. an adenovirus). Once inside, it is transcribed to *HSV1-tk* mRNA and then translated on the ribosomes to the protein (enzyme) HSV1-TK. A radio-labelled reporter probe (e.g. 8[^{18}F]-fluoroganciclovir – [^{18}F]FGCV) is also introduced into the cell, where it is phosphorylated by the kinase HSV1-TK (which is the product

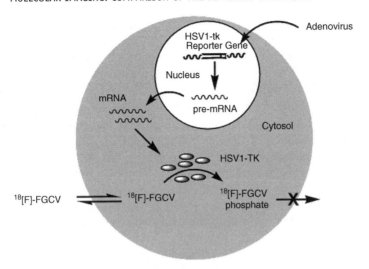

Figure 13.5 Schematic representation for imaging Herpes Simplex virus type 1–thymidine kinase (*HSV1-tk*) reporter gene using an ^{18}F-labelled probe

resulting from the gene incorporated in the cell). Once phosphorylated, [^{18}F]FGCV cannot easily travel across the membrane and as a consequence it is trapped within the cell. The amount of radio-labelled probe present in the cell is, hence, proportional to the level of HSV1-TK enzyme activity and directly reflects the expression of the *HSV1-tk* gene (Figure 13.5).

A second general approach which is currently being developed to image gene expression is by labelling short oligodeoxynucleotides (typically 12–35 nucleobases long) and studying their interactions with the mRNA transcribed from a particular gene. The modified radiolabeled antisense oligodeoxynucleotides (RASON) is prepared so that it is complementary to a small segment of the target mRNA (associated to a specific gene) and the imaging occurs at the transcription level. Messenger RNA molecules are excellent targets since they can form very strong complementary bonds (via hydrogen bonding) with specific antisense DNA fragments.

Initially, RASON probes were prepared using γ-emitting ^{111}In and ^{99}mTc-labelled antisense oligodeoxynucleotides for use in single photon emission computed tomography (SPECT) imaging. However, the problem with this type of labelling is that the probes are usually relatively large (a chelating ligand is required to complex the metal centre) and hence can interfere with the recognition between the RASON and the targeted mRNA. As a consequence, there has been recent interest in preparing ^{18}F-labelled oligodeoxynucleotides for PET imaging. Recently, an 18-mer oligodeoxynucleotides labelled at the 3′ end with ^{18}F has been prepared and its biodistribution *in vivo* has been studied. Although this RASON was not prepared to target a specific mRNA, the investigation has provided important information on the pharmacokinetics of the probes (besides demonstrating that the labelling does not affect the biodistribution of the oligodeoxynucleotides).

In spite of the enormous impact that radiolabelled antisense oligodeoxynucleotides could have in the study of gene expression, there are still a limited number of successful examples where it has been employed (and most of the cases have been employing SPECT rather than PET imaging). Several reasons account for this relatively slow development of a powerful approach: (a) the number of target molecules per cell is low and hence sensitivity is an issue (it is estimated that in a given cell the number of molecules of mRNA typically ranges between a few hundred and a few thousand while, for example, proteins are present in the range of thousands to millions); (b) biological delivery barriers prevent the probe from reaching the target; and (c) degradation of the labelled nucleotides by nucleases occurs *in vivo* (although some recent studies have demonstrated that it is possible to use more robust alternatives to the oligodeoxynucleotides as the targetting moiety).

13.3.4   Molecular imaging in drug discovery and development

The development of new drugs is a complex, expensive (it is calculated that the development of a new drug costs up to US$800 million) and lengthy process (in average a drug takes approximately 12 years from discovery to approval). Consequently there is enormous interest in finding more efficient ways to discover and develop new drugs, from the identification of the targets through to testing and evaluating the drug candidates (in animal models first and eventually in humans).

Over the past few years advances in biomedical and chemical sciences have led to the development of more precise therapeutics aimed at specific molecular targets that are associated with disease. In order to understand and improve these directed therapies, it is necessary to have good means of visualizing the effects that a drug–target interaction has on the process associated with a specific disease. This would provide important information whether a specific drug has reached its target or not, if it affects the function of the targetted species and if the drug–target interaction has an effect on the course of the disease. On the other hand, for the successful development of novel therapeutic agents it is essential to study the way in which the body absorbs, distributes, metabolizes and excretes them (i.e. the pharmacokinetics and pharmacodynamics of the drug). Molecular imaging – particularly PET – is playing an increasingly important role in all of these aspects of drug discovery and development.

It has already been shown in preceding sections of this chapter that PET is a useful tool to *interrogate* biological processes at a molecular level, providing important clues on how, for example, small molecules interact with enzymes or biological receptors. Although the use of PET in the discovery and development of novel therapeutic agents is still relatively limited, it has already been successfully employed in several receptor occupancy studies, either at a single time point or studying the drug action as a function of time. For example, radiolabelled species have been used in clinical studies for the development of new drugs aimed to treat diseases associated with the central nervous system, in particular by establishing the relationship between drug dose and the occupancy of target sites in the brain (either in a receptor, the active site of an enzyme or a transporter binding site). In order to carry out these studies, radiotracers with high affinity and selectivity for the targeted binding sites are synthesized. Once administered, they bind to the targeted receptor and hence provide

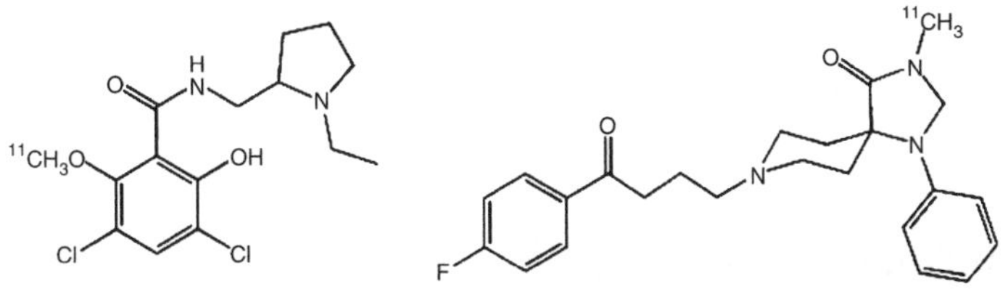

Figure 13.6 Structures of two radiotracers used for occupancy binding studies in the central nervous system

tomographic images of the site of interest. *In vivo* competition studies can then be carried out by adding the unlabelled drug candidate and analysing the PET images of the site of interest; this can then provide a direct measure of the occupancy of the specific receptor in the presence of the drug candidate. A specific example of this approach has been used to show that drugs currently used for the treatment of schizophrenia block certain post-synaptic dopamine receptors (see Section 13.3.2) with occupancies higher than 70 per cent at therapeutic doses. The radiotracers that have been used for these studies have been either N-[^{11}C]-methylspiperone or [^{11}C]raclopride (Figure 13.6).

To date, there are several other available PET radiotracers that are known to bind strongly and selectively with a range of receptors in the brain. Hence, it is possible to screen new drugs and evaluate if they interact with these receptors or not. If new therapeutic agents are developed which interact with other receptors (for which no radiotracer is yet known), then the process is more complex since it involves not only synthesizing the new candidate drugs, but also the corresponding radiotracer to image it. Sometimes, the drug candidate itself can be labelled and used as a radio-ligand.

It has already been pointed out that in drug discovery and development it is also very important to be able to study the absorption and metabolism of the therapeutic agent in the body. Although most of the pharmacokinetic and pharmacodynamic studies needed for clinical approval of a drug are first carried out in animal models, there is increasing interest in doing such studies directly in humans since this would reduce considerably the failure rate associated to the inappropriate human metabolism of drug candidates. To tackle this, a new method known as microdosing has been developed in which human metabolism data on new potential drugs can be obtained. This approach comprises administration of sub-pharmacological or sub-therapeutic dose of a novel drug candidate to a human volunteer. The potential drug can be labelled with a radioisotope and hence PET can be used to monitor its absorption and metabolism.

Some studies have already been successfully carried out using PET microdosing. For example, valuable pharmacokinetic data has been obtained by studying the distribution of the anticancer labelled agent ^{11}C-N-[2-(dimethylamino) ethyl]acridine-4-carboximide (DACA) in cancer patients. In the study, ^{11}C-DACA was administered at one-thousandth of the proposed starting dose for a conventional phase I study but still provided important information on distribution of the drug in various tumour and normal tissues. The authors concluded that there was great potential importance of this type of study in predicting the activity and toxicity of drugs in their early development.

13.4 Conclusions and outlook

The past two decades have seen important developments in PET-based molecular imaging *in vivo*. This technique (jointly with several others such as MRI, SPECT and optical imaging) is being instrumental in gaining a better understanding of biological processes at a molecular and metabolic level, besides having an impact on our way of diagnosing, preventing and treating diseases. As has been shown throughout this chapter, we now have the possibility of measuring by noninvasive techniques a variety of events such as the distribution of enzymes, how genes are expressed and the interactions between receptors and their targets.

Thus far, the great developments achieved in molecular imaging are likely to be only the 'tip of the iceberg'. This highly multidisciplinary area of research is expected to continue developing steadily over the next few years. For example, it is expected that the refinement of current chemical synthetic methodologies will produce more sophisticated probes with higher selectivity and sensitivity. On the other hand, advances in detector technology and image reconstructing techniques for PET are likely to generate a new generation of instruments with better spatial resolution and sensitivity. Novel approaches to targetting specific tissues to be imaged and new methods to deliver the probes (e.g. more robust vectors for gene transfer) will also need to be developed. Finally, there is an increasing interest in designing multiple-imaging probes that can make use of two or more imaging modalities simultaneously. This will allow the advantages of each specific imaging techniques to be played up to improve our understanding of biological processes at a molecular level.

Further reading

For general overviews on molecular imaging see:

Gillies, R.J. (2002). *In vivo* molecular imaging. *J. Cell. Biochem. Suppl.* **39**: 231–238.

Herschman, H.R. (2003). Molecular imaging: looking at problems, seeing solutions. *Science* **302**: 605–608.

General references on PET:

Phelps, M.E. (2000). Positron emission tomography provides molecular imaging of biological processes. *Proc. Natl Acad. Sci. USA* **97**: 9226–9233.

Hnatowich, D.J. (2002). Observations on the role of nuclear medicine in molecular imaging. *J. Cell. Biochem. Suppl.* **39**: 18–24.

Phelps, M.E. (2004). *PET: molecular Imaging and its Biological Applications*. Springer: Berlin.

For an overview of PET instrumentation:

Turkington, T.G. (2001). Introduction to PET instrumentation. *J. Nucl. Med. Technicl.* **29**: 4–11.
For applications of PET to drug discovery see:

Rudin, M. and Weissleder, R. (2003). Molecular imaging in drug discovery and development. *Nat. Rev. Drug Discov.* **2**: 123–131.
Lappin, G. and Garner, R.C. (2003). Big physics, small doses: the use of AMS and PET in human microdosing of development drugs. *Nat. Rev. Drug Discov.* **2**: 233–240.
References providing an overview of PET in imaging gene expression:

Gambhir, S.S., Barrio, J.R. Herschman, H.R. and Phelps, M.E. (1999). Imaging gene expression: principles and assays. *Top. Mol. Biol.* **6**: 219–233.
Min, J.J. and Gambhir, S.S. (2004). Gene therapy progress and prospects: non-invasive imaging of gene therapy in living subjects. *Gene Ther.* **11**: 115–125.

14

Chemical genetics

Piers Gaffney

Imperial College, London, UK

14.1 Introduction

Although the term 'chemical genetics' has been sweepingly used to describe a wide range of studies bringing together the interactions of small molecules with bio-molecules under a genomic umbrella, its aim can largely be defined as the search for chemical agents that act as conditional switches for gene products, either inducing or suppressing a state or phenotype of interest. Hence, chemical genetics can be seen as an offshoot of functional genomics, and it also feeds into the rapidly growing post-genomic field of systems biology. The ultimate ambition would be for each and every possible protein interaction, to find one small molecule that specifically disrupted it, and for each association to be targetted by a different small molecular. Although this would be a mighty biological tool, and much will be gained by applying the principals on a limited scale, it is probably unachievable on an entire genome.

The concept of chemical genetics seems to have come about for two reasons. First, the probing of biological systems with chemical entities has been re-visited in the light of all the newly available genomic information. Second, chemical genetics can be encapsulated within a critical new perspective: the selective interaction of a small molecule with a protein may be regarded as functionally equivalent to mutation of that protein. As with classical genetics, chemical genetics is not an end in itself but a tool for learning about the functions of biological macromolecules. So, although the experimental protocols that are now being developed under the chemical genetics banner are not completely novel, the results can be cross-referenced with genomic data sets, and fed back into further rounds of study if necessary, to generate a new understanding of the larger system.

Chemical Biology   Edited by Banafshé Larijani, Colin. A. Rosser and Rudiger Woscholski
© 2006 John Wiley & Sons, Ltd

14.2   Why chemicals?

Despite the functional equivalence of mutating a protein with perturbation by a chemical, treating an organism with a chemical has several fundamental differences to conventional genetics. Most significantly, although it seems obvious, at least for *forward* chemical genetics (see below), one does not need a genetically modified organism, i.e. a chemical can be added to the wild-type organism, and it is easy to add one chemical to several cell lines, but time-consuming to genetically engineer each cell line. This is particularly true for mammals where applying recombinant techniques is especially difficult because of their relatively large, diploid genomes and slow rates of reproduction.

The difficulties of genetic engineering are greater when a *conditional* mutation is required. However, addition of a chemical may be equated with activation of a conditional mutation. It is common that one gene product can have different functions at various stages of an organism's lifecycle. These separate roles of the same protein are difficult to disentangle genetically, but it is trivial if a chemical genetic equivalent of that mutation is available. Furthermore, mutagenization often induces lesions that are fatal in early embryonic development, making such mutants difficult to study genetically. In this case the innately conditional chemical genetic approach makes study of the protein's function much more practical. It also permits study of any effects that the protein may have later in development, avoiding the fatal consequences of early disruption.

he next most telling difference between classical and chemical genetics is the time-scale. Instead of simple deletions, more difficult temperature-sensitive mutants, protein expression controlled by an antibiotic resistance gene promoter, or some other conditional system are usually preferred. Whilst such techniques are powerful, they all have relatively long induction times that prevent them being used to study faster processes (<minutes to hours). However, chemicals can diffuse into a cell in a matter of seconds. This is much faster than any genetic experiment, and lends itself to current interests in intracellular signalling where events frequently occur on short time-scales.

A significant drawback of mutating a protein in an apparently critical pathway is that it often leads to complementary mutations elsewhere in the genome that compensate for the damaged system by up-regulation of related proteins; this mechanism for side stepping a mutation's effects may be very common. In the chemical genetic approach, there can be no selection through evolutionary bottle-necks, leading to compensation of the disrupted protein's function. There may also simply be redundancy that allows a damaged system to be circumvented, but not within the duration of a chemical genetic experiment.

Most commonly genetic strategies are used to either delete a protein or completely disrupt its cellular activity, so the effects of these mutations are usually quite black and white. Even taking into account over-expression and heterozygous mutants, there is no mechanism by which to truly graduate the response. However, a chemical's effects depend on the added concentration, so the experiment is completely tunable. Additionally, if the chemical is washed off the experiment, its action is often rapidly reversible, allowing cells to recover from exposure, again impossible in many genetic approaches.

Finally, many proteins have multiple functions and, particularly in the case of signalling, they often act as a scaffold for other proteins to assemble around, integrating the signal from multiple binding partners. Complete loss of a protein, or a significant part of it, in a deletion mutant, aimed at the function of one domain, often unintentionally disrupts the roles of all the other domains. However, chemical modulation of a protein is localized to the chemical's binding site, usually affecting only a small part of a domain of a large protein, or even just one conformation of a dynamic entity, without interfering with any other potential active sites. Thus, a chemical genetic approach can in some ways be more precise than a genetic one.

The most significant downside of chemical entities as probes for biology is that the approach is not so far generalizable in quite the same way that genetic manipulation is. So, although it may well be that a probe can eventually be found for any protein function, there are so few currently known that genome-wide studies using small molecule probes – chemical genomics – are not yet possible.

14.3 Chemical genetics – why now?

For several decades the biological sciences have been dominated by genetics and the corresponding techniques of molecular biology. These techniques substantially supplanted earlier strategies for perturbing biological systems with small chemical agents because it was very time-consuming to establish the bio-molecular targets of such compounds. By contrast, the logical force behind genetics was the precision of selectively altering just one base-pair in an entire genome – roughly 3 billion base-pairs for humans. Its power derives from relating proteins to known genes, then having the ability to totally specifically adjust any process the protein is involved in by mutation and studying the mutation's effect in a living organism.

To make maximum use of genetic strategies for probing biological systems, whole genomes have been sequenced. With hundreds of genomes now in hand, the aim has recently become to obtain a more holistic interpretation of complex cellular processes. Although the human genome is less than twice the size of those of worms and flies, and only five times larger than that of unicellular yeast, we seem so much more complex. Since this difference from bugs to man correlates poorly with the increased number of genes, it has instead been interpreted as a multiplication of the amount of crosstalk between genes and between gene products. It is the aim of the rapidly growing field of systems biology to characterize and quantify the feedback between control networks. The emergence of chemical genetics complements the genetic approach for probing these networks.

In several ways present-day chemical genetics is similar to drug discovery screening processes prior to the 1970s. This changed with the advent of recombinant DNA technology and the ready availability of artificial expression systems, when drug discovery switched to screening compound libraries against purified single proteins as rational disease targets. Implicit in this approach to drug discovery is the assumption that the purified protein is directly in a pathway that controls the rate-limiting step underlying a given phenotype. However, when a compound identified by this strategy proceeds to *in vitro* testing, or even medicinal use, it is virtually impossible to rule out biologically significant interactions with bio-molecules belonging to entirely different cellular systems. Indeed, it has emerged that the success of several compounds was ultimately dependent on their 'off-target' inter-actions. Perhaps as a consequence, this mechanism-based approach has only yielded a limited crop of drug hits.

This uncertain transfer of compound specificity from the *in vitro* to the *in vivo* environment has partly driven search strategies back to phenotypic screens. Pre-viously, the uncertain specificity of any chemical's bio-molecular associations had been seen as their Achilles' heel. However, when screening on the basis of phenotype, the absolute specificity of a chemical's interaction with its target is not strictly required at its active concentration. This is because either off-target interactions are below their active level – otherwise they would generate a new phenotype – or at least any off-target interactions cannot be on the main biochemical pathway corresponding to the phenotype. Furthermore, phenotypic screening only provides a hit when a chemical interacts with an unavoidable check-point in a given biochemical pathway. This reversion to phenotypic screening has led Stuart Schreiber, one of the godfathers of modern biological chemistry, to describe chemical genetics as 'a return to the good old days' of drug-candidate discovery.

14.4 The relationship between classical genetics and chemical genetics

Classical genetic strategies are divided into forward and reverse approaches (Table 14.1). In forward genetics the aim is to identify what genes underlie a particular behaviour. Thus an organism under study is first randomly mutagenized. A phenotype of interest is then selected from amongst a large collection of mutants. Finally, mapping strategies are used to locate the underlying gene that has been mutated, ultimately leading to a biochemical interpretation of the phenotype. In reverse genetics the aim is to figure out the biological significance of a given protein. So the gene corresponding to a protein of interest is first mutated and expressed in the study organism. Any phenotypic change from the wild-type that is observed is then used to interpret the normal function of the protein.

There are chemical genetic parallels of both forward and reverse genetics, combining many aspects of biology, chemistry, pharmacology and medicine. In forward chemical genetics the study organism is screened against a large number of chemical entities. Molecules that induce interesting responses are identified – this is equivalent to observing a novel phenotype in forward genetics. The most difficult step is then to locate the cellular target with which a given small molecule associates. This is often, but need not be, a protein. Chemical genetics in the forward direction, which

Table 14.1

Forward genetics		Reverse genetics		
Classical	Chemical	Classical	Protein–ligand engineering	Chemical
Random mutagenization of organism	Add chemical library to organism	Mutate gene of interest	Mutate gene of interest and modify known ligand to fit mutant protein	Add chemical library to purified protein of interest
Select mutants with phenotype of interest	Select molecule that gives phenotype of interest	Insert mutant gene into organism	Insert mutant gene into organism	Identify molecule that binds to the protein
Identify mutated gene	Identify protein that binds the small molecule	Observe effects of mutation on phenotype	Add modified ligand to engineered organism, observe phenotype	Add molecule to organism, observe phenotype

Adapted from Stockwell, B.R., with permission from *Nature Reviews Genetics* (200; **1**: 116), Macmillan Magazines Ltd.

constitutes the bulk of the rest of this chapter (Section 14.5), can clearly be seen as a branch of functional genomics and so has attracted much recent interest.

The logical reversal of this process would be to screen an isolated protein against libraries of compounds, until a specific association is identified – the parallel of creating a mutant of a known protein in reverse genetics. A study organism would then be exposed to this chemical and any response assumed to correspond to disruption of the selected protein's normal function. Whilst this clearly describes reverse chemical genetics in the strictest sense, this is very similar to classical rational protein targeting as used in drug hunting (Section 14.3), although a chemical geneticist would have a broader interest in the phenotypic implications of the selected chemical than a drug hunter. Consequently, what is now often referred to as reverse chemical genetics is a combination of chemicals with reverse genetics. Although such an approach is already familiar to us and is known as protein–ligand or ligand–receptor engineering, it is the recent setting of these concepts within the larger genomic arena that has broadened their appeal to the wider biological community. Consequently, only a relatively brief discussion of reverse chemical genetics is given in Section 14.6.

14.5 Forward chemical genetics

A forward chemical genetic screen is used to identify compounds on the basis of a phenotype that is assumed to correlate with an underlying cellular process. Once a hit has been obtained, this small molecule is used as a probe to identify the target of its action. With this information in hand, the small molecule can be added to an experimental organism to disrupt the newly identified pathway at will, and can be used in further studies in the same way as a conditional mutation. This process implies three basic requirements for a forward chemical genetic screen:

- first, obtain a library of compounds;

- then, screen the compounds for a change of phenotype;

- finally, identify the bio-molecular targets.

The issues surrounding these basics are examined in greater detail in Sections 14.5.1–14.5.3.

14.5.1   Obtain a compound library

To conduct a chemical genetic phenotypic screen there is an absolute requirement for a large selection of bioactive molecules. The primary limits on these molecules' availability are the synthetic routes to them. For this reason there are three main classes of such compounds; natural products, synthetic peptides and pharmaceutical-like molecules.

Natural products   Natural products are simply chemicals isolated from a large variety of wild organisms. They are usually secondary metabolites that typically participate in an organism's defence against predation or pathogens. Historically, when a plant, animal, marine organism, fungus or bacterium displayed unusual toxicological, pathological or other bioactive effects, a curious scientist would harvest and process the organism using all sorts of physical techniques to try to extract the active component. It was plain bad luck if the desired property depended on the synergistic action of two components in a mixture.

In general, natural products have molecular weights significantly larger than 500 and the elements carbon, hydrogen and oxygen predominate, with little nitrogen and less of other elements. The size of natural products provides sufficient contact area for them to develop many small interactions with their targets, the sum of which contributes to their highly valued discriminatory powers. This large contact area also makes them the reagents of choice to interfere with protein–protein contacts.

Natural products commonly display complex architectures that have evolved through countless iterations of the Darwinian arms race to hone their functions. Despite the difficulties, their powerful bioactive effects have driven long and costly research programmes to develop synthetic routes to them. In particular, natural products possess features that lock their structures into pre-organized geometries (especially multiple rings of atoms) that are commensurate with the constellation of features in their target binding sites – compared with a more flexible structure, this minimizes the loss of entropy on association that would otherwise favour dissociation. In addition, many natural products have masked reactive groups that are only activated in their target binding sites.

Box 1   Solid-phase peptide synthesis

When synthetic peptides are required in any more than tens of milligram quantities they are prepared in standard chemical laboratory glassware using standard organic synthetic techniques. However, for small amounts of peptide the versatile technique of solid-phase synthesis is used. Here the first residue of a desired sequence is covalently attached to an inert solid support, usually in the form of small beads with typical diameters of 10–200 μm and a capacity of 1–1000 pmol. The chain is extended one residue at a time by immersing the beads in a cycle of the appropriate reagents, and the reactions are driven to high conversions by large excesses of reagents.

Almost always, the C-terminal of the synthetic peptide is attached to the solid support and the N-terminal free $-NH_2$ is exposed at the growing end of the peptide. During chain extension, a condensation reaction is performed between the free amine and a carboxylic acid, $-CO_2H$, of the next amino acid residue to form an amide bond. As the condensing agents used for this key reaction would also alter many amino acid side chains, these additional reactive groups are usually protected until the chain has reached its full length. Additionally, the amine of the incoming residue must also be protected to avoid random polymerization. The coupling cycle consists of (i) condensation of a dissolved amino acid carboxyl group with the N-terminal of surface-tethered poly-peptide chain; (ii) capping of any residual unreacted amines (this simplifies final purification); and (iii) N-deprotection of the new N-terminal amino acid. There are two complimentary sets of synthetic reagents, most obviously differing in their monomer N-protective groups (Fmoc or Boc), although the consequences of this selection greatly affect many aspects of the subsequent synthesis.

A particularly convenient feature of this technique, by contrast with solution-phase synthesis, is that there is no need for multistep purification of the product because the by-products and debris are easily washed away from the solid support before proceeding to the next coupling cycle with a new amino acid. The solid support can be reacted with reagents in various types of apparatus; it can be simply held in a tube sandwiched between porous frits whilst reagents are pumped past, or reagents can be injected into and sucked out of vials sealed with a rubber septum, or it can be treated in 96-well plate format blocks of miniature reactors. There are also several proprietary styles of solid support designed for convenience of use.

Despite the difficulties, the powerful bio-active effects of natural products have driven long and costly research programmes to develop synthetic routes to them. However, of the may so far devised, few readily adapt to the preparation of compound libraries. Also, although there are very many natural products known with established protein specificities upon which to model a small library, the availability of an appropriately matched natural product to any given protein target is a matter of luck. In this regard, the frequent difficulty of extraction of these classes of compound from natural sources has made it difficult to assemble broader libraries to screen, although not impossible. For instance, after partial fragmentation of microbial DNA from soil samples, some of which will contain entire operons for secondary metabolite synthesis, insertion of this material into bacterial artificial chromosomes can be used to generate a natural product library.

Peptides The most obvious feature that makes synthetic peptides good candidates for library synthesis is that, since nature is full of them, they are clearly bioactive. However, the delivery properties of peptides vary widely, and they are subject to enzymatic degradation, giving them potentially short half-lives *in vivo*. In spite of these limitations, they lend themselves to library synthesis because, being a linear hetero-polymer, they are amenable to automated synthesis (Box 1), and therefore for the preparation of libraries.

Two distinct strategies of peptide library preparation are commonly used; parallel synthesis and so-called split-and-mix chemistry (also known as divide–couple–recombine, portion-mixing and split-pool synthesis). Both are often described as combinatorial chemistry (combichem), although this is only strictly true for split-and-mix synthesis. Parallel synthesis is simply the preparation of many batches of peptide at the same time in separate *parallel* channels of one or more machines. In the case of library synthesis, the amino acid sequence is systematically varied between channels. Commonly this allows the synthesis of several hundred different sequences at the same time. It has the advantages that larger amounts are usually prepared than in split-and-mix, and that one can easily tell what sequence of residues was used in any potential hit.

The split-and-mix strategy appears deceptively similar to parallel synthesis, but can produce libraries of several hundred thousand compounds in one run. Although it can use the same style of apparatus as for parallel synthesis, there would normally be as many synthesis channels as one has monomers – 20 in the case of natural amino acids (although cysteine is often omitted). However, after the first coupling cycle has been completed, the batches of synthesis beads from each channel would all be combined, mixed well, and re-divided back into the 20 synthesis channels – the step from which the method takes its name. After this, another round of chain extension would commence. In this way all possible sequences are prepared at the same time, but each bead of solid support only contains one sequence. As the number of compounds rises exponentially with each chain extension cycle (hence it is *combinatorial*), large numbers of monomers limit the number of cycles that can be performed – usually to four rounds (i.e. a tetrapeptide library) with 20 different monomers, $\equiv 20^4 = 160\,000$ compounds. This is because the number of possible sequences should be significantly less than the number of synthesis beads that can be contained in a reasonable volume, to ensure that all possible sequences are actually prepared. One downside of this strategy is that you only get one bead's worth of compound to screen, usually whilst still attached to the bead. Also the identity of any hits will have to be extracted by chemical analysis of the bead's contents. In the case of peptide synthesis, sequence analysis can be performed by Edman degradation. However, for libraries of other compounds this can be more difficult, sometimes being assisted by the inclusion of a chemical barcode to be used as a more easily decipherable record of a bead's synthetic history.

Drug-like Molecules Drugs and pharmaceutical-like compounds, often referred to as pharmacophores, are quite different again from the first two classes of library compounds. Generally they are smaller than natural products and, whilst they do have some overlap, they have more uniform physical properties. Also, they do not have the stability problems of peptides and are effectively designed for ease of delivery. This is because they are usually made to conform to Lipinski's 'rule of five', so-called because the four numerical values to which the majority of drugs subscribe are either close to, or a multiple of, five. Thus, there should be fewer than five hydrogen bond donors, the molecular weight should be lower than 500, a constant defining a drug's greasiness (logarithm of the partition constant between octanol and water, log$P$) should be less than 5, and there should be fewer than 10 hydrogen bond acceptors. To a large degree these constraints relate to a molecule's ability to cross biological membranes by passive transport, hence their cell deliverability, and provide a guide to their pharmacokinetics. The design limitations imposed by these rules can be ignored if there is a biological transport system that assists a chemical's access to the intracellular environment, as is often the case for natural products.

Beyond these general properties the precise structures of drug-like molecules vary enormously. Even so, several other common features can be identified. Pharmacophores do not contain any obviously reactive or toxic groups. Also, they rarely contain any carbon atoms with four different groups covalently attached around them. This type of substitution prevents a molecule from having any symmetry, so that two versions of the same molecule can be isolated differing only by being mirror images of each other. These right- and left-handed forms, or enantiomers, are chemically identical in their reactions with simple reagents but can bind totally differently to complex bio-molecules. This is because enzymes and biological binding sites possess the same handedness, or chirality. However, only one of the two possible mirror images is produced by nature so binding of a small molecule to a bio-molecule almost always distinguishes one enantiomer from the other. Since separating or generating chirality synthetically is often nontrivial, and it may be difficult to ensure that a chiral compound is completely free of its potentially bioactive enatiomer, chirality is only usually included in the chemical synthesis of a pharmacophore if it is essential for high bioactivity.

Pharmacophores usually address the tightly confined environments of enzyme active sites or protein binding sites for endogenous small molecule ligands. Their limited size means that they are prone to recognizing the active sites of proteins related to their actual target because they have insufficient reach to discriminate between related proteins only having dissimilar features distant from the binding site. Also they have previously been considered too small to disrupt the large number of low-energy contacts present in protein–protein docking sites, which at about 800 Å^2 are too large an area for a molecule of only 500 Da to be able to cover. Despite this, the recent discovery of much smaller hot-spot residues within protein–protein contact footprints, and allosteric structural changes during protein–protein binding, both of which provide much smaller sites for pharmacophores to address, suggest that many docking sites are amenable to attack.

Library design, synthesis and diversity   Irrespective of the chemical genetic para-digm, much effort has been dedicated in the past decade to the preparation and screening of chemical libraries. They continue to play an essential role in classical drug discovery, both in academe and industry, and the contents of some are available commercially. The total number of possible chemical structures that could be assembled from just carbon, hydrogen, nitrogen and oxygen with molecular weights under 500 is incomprehensibly vast. As a result a library of even a few hundred thousand compounds is hardly a drop in this ocean. Furthermore, biological systems use an incredibly small fraction of all possible molecular diversity, with perhaps just a few hundred basic structural classes. Thus, to search the right portion of biologically relevant chemical structure space, and so have a reasonable chance of a hit, will depend on a careful choice of library. However, the underlying principals that characterize biological chemical structure space are still poorly defined, and so the principal constraint on library design is often synthetic dexterity.

There is a spectrum of approaches to library design and synthesis. At one end there is classical synthesis of a specific target, or of a small number of variations upon a theme, as in medicinal chemistry. This has been dubbed target-oriented synthesis, TOS. In this case there is rational selection of all or most of the atomic groupings of the target based on prior knowledge; for instance, detailed spectroscopy of an unusual natural product, or a structure–activity map of an active site plus shared features with other known agonists/antagonists. A systematic plan for the molecule's synthesis is then developed, working back from the complex final target by logically cutting it into smaller simpler pieces at known sites for established reactions, so generating a retrosynthetic analysis. The analysis is then reversed to provide a synthetic plan for a specific molecule and some close relatives. Although this is the way that pharma-ceutical companies have built up their collections over many years, it cannot provide newcomers to the field with a library.

Then there is focussed library synthesis, FLS. In its broadest sense this involves synthesis of a larger number of compounds than TOS, often using solid-phase chemistry (Box 1), but where all the products are still quite similar. FLS allows more exhaustive exploration than TOS of a relatively small volume of biological chemical structure space. Peptide library synthesis can be considered one of the first examples of this category because an underlying polypeptide backbone, which is the same for all members of the library, is elaborated with variable side-chains. Taking its lead from peptides and oligo-nucleotides, much effort has been directed over the last decade towards library synthesis with substrates bound to solid supports, to facilitate the generation of large numbers of different compounds in parallel or combinatorial syntheses. This can be substantially more difficult than for peptide synthesis because of the need to include a much wider range of reactive reagents to prepare new functional groups and structures.

Generally in FLS an underlying skeleton is modified by the attachment of variable appendages to one or more sites. The majority of commercial libraries are of this sort and tend to contain drug-like molecules. It has also become common to take the framework of a natural product, or substructures of it, as a starting point for this class of library. This is founded on the hypothesis that such structures have been refined by evolution for their utility, and so have a biologically privileged architecture. Indeed nature has been known to do much the same as synthetic chemists (for instance, with guanacastepenes) by using redox reactions to re-model a basic core to generate a suite of diverse molecular architectures. However, this approach has not so far been entirely successful. Although it provides a large number of compounds, with different formulae and structures, they do not explore a very wide volume of chemical space because the diversity-imparting variable appendages are always presented at the same points in three-dimensional space defined by the underlying molecular skeleton. Thus, as far as the biological targets are concerned, there is not much chemical diversity, just a large number of similar compounds.

Realizing this limitation, a new aim has emerged to produce combinatorial synthetic libraries of compounds with complexity approaching that of natural products, an approach that has become known as diversity-orientated synthesis, DOS. For this the core molecular skeletons of the library must also be varied to achieve a broader search of biological chemical structure space. In this context, reactions that construct complex skeletons in one step have proved very versatile; the best known of these is probably the Ugi four-component reaction. This assembles an amine, a carboxylic acid, an aldehyde and an isocyanide, the side-chains of which can all be varied to greater or lesser extent, into a dipeptide-like core. Another key set of reactions forms rings of atoms. These can generate several points of stereochemical complexity in one step, as well as reducing the floppiness of the skeleton (cf. natural products, above). Examples of these are the Diels–Alder cyclization of a diene and an alkene to give cyclohexene derivatives, and the ring opening–closing metathesis reaction, where one molecule of ethylene is ejected in the process of joining two alkenes, $R\text{-}CH{=}CH_2$, together, $\rightarrow R\text{-}CH{=}CH\text{-}R + CH_2{=}CH_2$.

The above techniques for construction of the skeleton, and many more, have been combined with later variation of appendages in a large number of FLS strategies. However, to expand the search volume of DOS beyond that of FLS, greater complexity has to be generated. Currently approaches are being explored that use two complexity-generating reactions in succession. The difficulty is how to ensure that all the products of the first reaction – by definition, already having a significant degree of stereochemical variety – are substrates for the second reaction. Also, to increase the diversity around a given skeleton, catalysts and conditions are being developed to override the stereochemical preferences of common reactions so that the directions in which appendages protrude from the core can be varied. The rationale for combining these reactions systematically in the design of a diversity-orientated synthesis is still being developed and, although there are some impressive individual reports, there is much yet to be done.

14.5.2 Screening

The term 'high-throughput screen' is often heard when searching for bioactive small molecules, but deserves closer scrutiny. It implies that a large number of compounds will be rapidly tested, but that few will be selected. Screens are best known from the pharmaceutical industry where both false positives and negatives are acceptable in the search for initial compounds that can be taken forward to more detailed studies. However, libraries can be chosen to probe the broader systemic properties of an organism when a global analysis of all the library members' effects is required, as is often the case in a chemical genetic screen. In this case error rates must necessarily be low for meaningful results to emerge, and observation at several concentrations would be desirable to assure the quality of the results.

In a forward chemical genetic screen the aim is to detect a change in phenotype triggered by a few members of a large library of compounds. For animals the time required to observe a phenotype, amounts of chemical required and ethics all militate against studies in whole organisms, although this is not the case with plants. Also, with a library of up to tens or hundreds of thousands of compounds, the assay will need to be miniaturized and automated, and therefore usually organized in the standardized 96-, 384- or even 1536-well format. Thus cell-based assays are usually the most practical approach. An optical read-out, or a simple colour-generating chemical assay, is then required that correlates with the desired phenotype, and that can detect suppression or enhancement on addition of a chemical. The *in vivo* forward chemical genetic screen also ensures cell permeability for a hit and an appropriate biological context, traits that cannot be assured from screening against a purified protein (as in simple reverse chemical genetics). However, it is more difficult to find a detectable molecular marker that acts as a signature for phenotypic modification than to detect protein binding in a cell-free system.

The marker for a phenotype is usually only an indirect indicator of the phenotype being observed. As in classical genetic screens it may require either prior molecular biology or the application of immuno-reagents to develop a detection system that can be automated. When the phenotype changes in response to an added chemical there would normally be significant changes to the overall protein expression profile of a cell. Consequently there will be many gene products that correlate with the new phenotype and can be used as a marker for it. If pathways associated with a desired phenotype are already known, a gene in this sub-group could be selected as the marker. Alternatively, if no such pathways are known, a DNA-chip (Box 2) could be used to search for proteins whose up- (or down-) regulation (i.e. mRNA concentration) is correlated with the phenotype of interest to serve as the marker. Having identified a marker, recombinant DNA technology can be used to insert a reporter gene into the study organism so that the gene product's expression is linked to the marker. The reporter protein then provides the detectable signal for a change of phenotype. For instance, the luciferase coding sequence could be linked to the promoter and enhancer elements of a marker gene making cells displaying the desired phenotype fluorescent. However, it is possible nowadays to omit the molecular biology and use the DNA-chip directly to provide a characteristic fingerprint of phenotypic modification, although this is still an expensive option.

Box 2   DNA chips

As the recent interest in obtaining an integrated overview of cellular behaviour has expanded, so has the search for massively parallel screening methodologies to interrogate their contents. Array technology is often the favoured approach to address such analytical demands because of its small scale, based around chemically modified surfaces having microscopic individual features, and rapid optical read-out of results. Such arrays rely on having a single, highly selective binding partner (*probe*) for each component of the mixture that one wishes to assay (*analyte*). The probes are bound to a surface in a grid pattern for which the probe identity at each point is known. Prior to assay a marker (usually a synthetic dye) is used to label all the potential analytes. On exposing the array to a solution of labelled analytes (e.g. a cell lysate), any analyte for which the corresponding probe has been included binds to the surface, when a microscopic fluorescent spot is detected on the array.

DNA arrays utilize this biopolymer's well-known ability to dimerize (the double helix) – single-stranded DNA can associate with a complementary single-stranded DNA (or RNA), the binding being entirely dependent on the linear sequence of DNA monomers (nucleotides). Many commercial companies will chemically synthesize any desired sequence of DNA, thus sufficiently long synthetic DNAs can be readily prepared to be cross-linked to arrays as probes.

DNA arrays are not usually used to detect DNA, but mRNA as a proxy for protein concentration in biological samples; the number of mRNA copies present at any one instant is often assumed to be approximately related to how much protein is being synthesized. However, although the concept is simple, this approach is renowned for false positives, i.e. the probes are (surprisingly) not completely selective. Part of the reason for this may lie in the amount of sample preparation required to add the label to the mRNA analyte: sample mRNA is copied, using reverse transcriptase, into complementary DNA (cDNA); this is then copied back into RNA, using RNA polymerase that is tolerant of nucleotide modification, so that chemically modified monomers that are either intrinsically fluorescent, or that fluorophores can be coupled to later, are included in the growing RNA. A notable feature of this procedure is that it provides signal amplification, but multiple steps can cause significant sample bias.

The problem with the preceding screens is that they only report changes at the transcriptional level. This is inappropriate if a phenotype is best mapped to a protein expression level or post-translational modification. In these cases a primary antibody can be raised against a protein marker of the phenotype. After growing up cells in a multiwell plate format and adding a single library member to each well, the cells are interrogated by fixing, permeabilizing and treating with the antibody to the protein marker. Any bound primary antibody is detected by a modification of the Western blot method.

The most general phenotypic screen is to detect morphological changes. However, to automate this requires modern image analysis and pattern recognition software. Even so, such screening has the added advantage of detecting related effects. So, for instance, when reduced motility is detected in a cell motility assay this could be due to the phenotype or cytotoxicity. Suitable software should rapidly eliminate the latter false positives.

14.5.3   Target identification

In forward chemical genetics, having found a compound from the library, a 'hit', that elicits the desired phenotype, the final step is to identify its bio-molecular point of action. This is usually the biggest hurdle, especially if the target is novel.

The first port of call would be to use classical biochemical methods. For instance, fractionated cell extracts could be tested for binding of a radio-labelled, or otherwise tagged version of the hit compound. Alternatively, the hit compound could be immobilized on a solid support and affinity chromatography used to detect its bio-molecular binding partner. This strategy can fail if the small molecule only has a relatively low affinity for its cellular target, or it may have a high affinity but the target protein has a low abundance.

These problems can sometimes be overcome with cDNA-based screens where protein concentrations are effectively boosted above their natural abundances, although all these methods require synthetic modification of the hit molecule. Probably the most developed technique is the phage display library where a virus that infects bacteria (a phage) is modified so that its genome contains DNA for the proteins to be screened fused to one of the virus coat particles. Thus, when a solution of phages, each displaying different coat fusion proteins is exposed to a surface with the hit molecule tethered to it, only those phages expressing a fusion protein of the hit molecule's target will stick. Once unbound phages have been washed away, bacteria are infected with the sticky phages, thereby greatly amplifying the target protein cargo which can then be much more easily identified.

There are two other technologies still in development with similar potentials. So-called yeast 'three hybrid screens' are related to the better known two-hybrid screens, where the binding interaction that is being hunted for triggers the assembly of a dimeric transcription factor. The functional transcription factor then stimulates expression of a reporter gene. One half of the transcription factor is fused to a known small-molecule receptor (say for a drug), whilst the other is fused to members of a cDNA library. When yeast colonies are treated with a synthetic drug-hit heterodimer, only those containing a fusion protein of the half transcription factor with the hit's target express the reporter gene. In a more brute-force approach, large-scale application of protein expression technology is used to obtain purified proteins covering a significant proportion of the proteome. These are printed onto micro-arrays, each spot containing one protein, and exposed to fluorescently tagged versions of the hit compound. The spot containing the target protein then fluoresces. The technology underlying these protein chips is somewhat similar to DNA-chips (Box 2), although the protein arrays are far less robust than their oligo-nucleotide counterparts.

The preceding techniques all require modification of the small molecule hit, but this can sometimes be difficult to do while still retaining their bioactivity. The final approach to identifying the hit's target is to use genetic screens that require no prior modification of the hit molecule. Thus, in the case of haplo-insufficiency where there is only one copy of the target protein's gene, correspondingly lower expression levels of this protein increase the cell's sensitivity to the hit compound. Inverting this principle, multiple gene copies should diminish a cell's sensitivity to the hit. Also, in a screen of deletion mutants only cells lacking the hit's target (or proteins closely linked to the pathway containing the target) will fail to show a change of phenotype on addition of the hit – this assumes that deletion mutants of the target are viable. Finally, one can screen on the basis of synthetic lethality. In classical genetics this is where mutation of either one of a pair of genes is viable, but the combination of both mutations at the same time is lethal. The pattern of lethality for treatment of single deletion mutants with the hit is compared with the synthetic lethal matrix. The aim is to find a pattern of lethality, caused by exposing single random mutants to the hit, which mimics the combination of a particular mutation with the same set of random mutations in the double deletion matrix. It is then likely that the hit's target is that mutant.

14.6   Reverse chemical genetics

In reverse genetics the aim is to start from a gene product (i.e. protein) of interest and determine what it does in a biological system. In the reverse chemical genetic variation, instead of using mutation to perturb the system, a small molecule is used. The kinds of chemicals suitable for this purpose are generally much the same as those described in Section 14.5.1 for forward chemical genetics for all the same reasons.

We are most familiar with this process in the context of drug hunting, when the aim is usually to find enzyme inhibitors or receptor agonists of a purified protein using screening *in vitro*, so-called target-based screening. Here too this is usually a labour-intensive task because chemical space is such a vast volume to search. Again, the choice between focussed vs diversity-orientated libraries must be made – the clearly defined search for pharmaceutical leads tends towards FLS, whilst academic chemical genetic screens for new pathways may require broader DOS libraries. Constraints might be created from analysis of a protein's structure–activity relationships with its natural substrates, or it may belong to a class of related proteins with known inhibitors that can serve as a lead for a focussed library, and an X-ray crystal or NMR structure may provide mechanistic insights. Furthermore, progress is now being made with virtual screening, where thousands of compounds are docked into a binding pocket *in silico*, allowing researchers to constrain the size of a later physical library screen. However, this is still in its infancy leading to unreliable predictions, partly because flexible torsions of both protein and small molecules are difficult to optimize computationally.

It is not the purpose of this review to expand on the early stages of drug discovery, even though this can be defined to lie within chemical genetics, firstly because it is not this aspect of the broader field that has attracted renewed interest from biological scientists, but also *in vitro* approaches all suffer from the inability to select for deliverability that naturally derives from an *in vivo* screen. One field that *in vitro* screens have rarely addressed, but for which *in vivo* reverse chemical genetics is generating new impetus, is in the disruption of protein–protein associations. The protein–protein interaction that one wishes to disrupt can be reconstructed in a yeast two-hybrid system, a technique that is familiar to many modern laboratories. Various reporter genes can then be used to detect when a hit compound disrupts the target interaction. However, once a hit is moved from the screen to application in a test organism, care must be taken to ensure that any observed change in phenotype is in fact a result of disruption of the expected protein–protein association.

Adopting a looser definition of chemical genetics, protein–ligand (or receptor–ligand) engineering is often included in the field of reverse chemical genetics (Table 14.1). Protein–ligand engineering is a combination of reverse genetics (i.e. a protein of choice is perturbed by mutation) with a synthetically modified small molecule ligand for the very precise manipulation of biological systems. Protein–ligand engineering is frequently used when an initial ligand is recognized by many proteins (not necessarily in a family), making the ligand's phenotypic effects difficult to correlate with a particular protein association. The ligand in this type of study could be a pharmacophore or a natural small molecule metabolite, e.g. a steroid or a nucleoside. The aim of this technique is most often to develop a modified protein–ligand pair in which a synthetic derivative of the original ligand binds specifically to a protein with a complimentary mutation, but the modified ligand is not recognized by the wild-type (WT) protein, i.e. the new ligand is orthogonal to the WT protein. The result is to reduce a many-to-one protein–ligand relationship to a one-to-one correlation. This simplifies interpretation of the new phenotype when the modified ligand is added to a study organism expressing the mutant protein. The simplest manifestation of this concept is the steric complementation strategy (Box 3).

For many applications it is also desirable that the mutant protein is orthogonal to the original small molecule ligand, i.e. the mutant protein cannot bind the unmodified ligand – the original and modified protein–ligand pairs are then said to be mutually orthogonal. Steric complementation does not usually ensure mutual orthogonality between the WT protein–ligand pair and the modified pair. For this engineering of polar interactions, such as hydrogen bonds or salt bridges, is usually required. This is effective because attractions between charges (or partial charges) are directional.

However, a broader assessment of the factors that must be traded off against each other in designing such a strategy are beyond the scope of this discussion.

Box 3   Protein–ligand engineering – steric complementation

Stuart Schreiber coined the term 'bump and hole' approach for this technique. A synthetic derivative of a known ligand for a protein is first prepared with an additional protrusion attached at a point situated in the interface between the ligand and its protein receptor. The protrusion or 'bump', which can be as simple as a methyl group, then prevents the modified ligand achieving proper contact with the protein so that the new ligand is no longer recognized. The aim is then to create a corresponding void in the binding pocket (the 'hole') to accept the bump, so allowing all the contacts from the original WT protein–ligand pair to resume and reconstitute a high-affinity association. In practice, a small library of bumpy ligands may be prepared and all of the members screened in a binding assay against point mutants of the receptor protein, with alanine replacing any residue in the vicinity of the ligand binding pocket. Notably, the bump and hole approach does not usually generate mutant proteins that discriminate against (are orthogonal to) the original ligand, which can just as easily occupy the expanded binding site as in the WT protein.

A good example of this approach is seen in the work of Shokat and co-workers on the viral v-Src kinase which can help to trigger cancer. They wished to establish the down-stream targets of this kinase using ^{32}P-ATP to radio-label its substrates. However, as perhaps 2 per cent of the human genome encodes for kinases, it would be very difficult to identify the products of v-Src kinase action in the cellular super-abundance of other kinases. Shokat and co-workers prepared 6-*N*-alkylated ATPs (A*TP) that no longer fitted the WT ATP binding site, then screened v-Src kinase mutants for those that could phosphorylate using an A*TP. Taking into account structural data, they were able to optimize an engineered v-Src/A*TP pair that had similar activity to the WT pair and that was not shut down by competitive inhibition from endogenous ATP. This technique has since been adopted by many other laboratories.

14.7   Closing remarks

Chemical genetics has the potential to offer insights into biological systems as powerful as, and somewhat complementary to, classical genetics. The use of cell-permeable small molecules to perturb the behaviour of biological macromolecules is a familiar tool to biological scientists, but the new perspectives generated by equating these interactions with mutations are changing the way they are used. For forward chemical genetics, identification of the cellular targets of small molecules is seen as the biggest hurdle to its application, somewhat paralleling the difficulty of locating the site of mutation in classical forward genetics. Even so, there are many methods for doing this, although none are ideal.

Luminaries in the field have highlighted the potential of chemical genetics, which may well be justified. However, to realize this potential requires fully integrated interdisciplinary scientific endeavour – sophisticated organic synthetic skills, a broad appreciation of the biological sciences, and investment in technologies such as robotics and image analysis. Whether institutions will sustain a commitment on this scale is questionable, particularly after the experience with many with the newer '-omics' sciences where simple concepts have promised so much, but spiralling technological costs and technical difficulties have squeezed many out of the field.

The greatest intellectual challenge for efficient application of chemical genetics is the choice of small molecule library. Although some libraries can be purchased, they are expensive and have been criticized for lacking true diversity. What is more, many conceivable studies would benefit from logical targetting of the library, but for this it is essential to have access to synthetic capabilities. It is for this reason that peptide libraries are still popular, even though their poor cell-permeability rules them out of many studies, because automated peptide synthesis technology is widely available and relatively user-friendly. However, the design and synthesis especially of diversity-orientated libraries is still very much in the chemical research laboratory.

The future utility of chemical genetics will depend upon access to libraries. Most current collections of diverse molecules are privately owned and unlikely ever to be widely available. Thus the future of this field probably lies in the hands of the synthetic chemists, hanging on their willingness to participate more deeply in interdisciplinary projects and develop custom libraries at affordable prices.

Further reading

Anson, L. (2004). Nature Insight, Chemical space. *Nature* **432**: 823–865.

Burke, M.D. and Schreiber, S.L. (2004). A planning strategy for diversity-oriented synthesis. *Angew. Chem. Int. Ed.* **43**: 46–58.

Spring, D.R. (2005). Chemical genetics to chemical genomics: small molecules offer big insights. *Chem. Soc. Rev.* **34**: 472–482.

Mayer, T.U. (2003). Chemical genetics: tailoring tools for cell biology. *Trends Cell Biol.* **13**: 270–277.

Koh, J.T. (2002). Engineering selectivity and discrimination into ligand–receptor interfaces. *Chem. Biol.* **9**: 17–23.

Stockwell, B.R. (2000). Chemical genetics: ligand-based discovery of gene function. *Nat. Rev. Genet.* **1**: 116–125.

Bishop, A., Buzko, O., Heyeck-Dumas, S., Jung, I., Kraybill, B., Liu, Y., Shah, K., Ulrich, S., Witucki, L., Yang, F., Zhang, C. and Shokat, K.M. (2000). Unnatural ligands for engineered proteins: new tools for chemical genetics. *Ann. Rev. Biophys. Biomol. Struct.* **29**: 577–606.

Index

Chemical Biology   Edited by Banafshé Larijani, Colin. A. Rosser and Rudiger Woscholski
© 2006 John Wiley & Sons, Ltd

This index was prepared by Neil Manley